CAMBRIDGE
UNIVERSITY PRESS

IT

for Cambridge International AS & A Level

PRACTICAL SKILLS WORKBOOK

David Waller

CAMBRIDGE
UNIVERSITY PRESS

University Printing House, Cambridge CB2 8BS, United Kingdom

One Liberty Plaza, 20th Floor, New York, NY 10006, USA

477 Williamstown Road, Port Melbourne, VIC 3207, Australia

314–321, 3rd Floor, Plot 3, Splendor Forum, Jasola District Centre, New Delhi – 110025, India

79 Anson Road, #06–04/06, Singapore 079906

Cambridge University Press is part of the University of Cambridge.
It furthers the University's mission by disseminating knowledge in the pursuit of education,
learning and research at the highest international levels of excellence.

www.cambridge.org
Information on this title: www.cambridge.org/9781108782562

First edition published 2017
Second edition published 2020

20 19 18 17 16 15 14 13 12 11 10 9 8 7 6 5 4 3 2 1

Printed in Poland by Opolgraf

A catalogue record for this publication is available from the British Library

ISBN 978-1-108-78256-2 Practical Skills Workbook Paperback with Digital Access (2 Years)

Additional resources for this publication at www.cambridge.org/9781108782562

....

....

Acknowledgments
The authors and publishers acknowledge the following sources of copyright material and are grateful
for the permissions granted. While every effort has been made, it has not always been possible to
identify the sources of all the material used, or to trace all copyright holders. If any omissions are
brought to our notice, we will be happy to include the appropriate acknowledgements on reprinting.

Cover image: adventtr/Getty Images

Photos and source file assets are reproduced by permission of the author.

Source files for Chapter 6 task 6 include photo by Big Foot Productions/Shutterstock, Chapter 7
task 6 include photos by INTERFOTO/Alamy Stock Photo, Maurice Savage/Getty Images, Stephen
Cooper/Alamy Stock Photo, Suljo/Getty Images, SSPL/Getty Images

DEDICATED TEACHER AWARDS

Teachers play an important part in shaping futures. Our Dedicated Teacher Awards recognise the hard work that teachers put in every day.

Thank you to everyone who nominated this year; we have been inspired and moved by all of your stories. Well done to all of our nominees for your dedication to learning and for inspiring the next generation of thinkers, leaders and innovators.

Congratulations to our incredible winner and finalists!

WINNER

Patricia Abril
New Cambridge School, Colombia

Stanley Manaay
Salvacion National High School, Philippines

Tiffany Cavanagh
Trident College Solwezi, Zambia

Helen Comerford
Lumen Christi Catholic College, Australia

John Nicko Coyoca
University of San Jose-Recoletos, Philippines

Meera Rangarajan
RBK International Academy, India

For more information about our dedicated teachers and their stories, go to
dedicatedteacher.cambridge.org

CAMBRIDGE UNIVERSITY PRESS

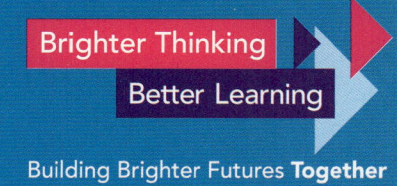

Brighter Thinking
Better Learning

Building Brighter Futures **Together**

〉 Contents

> How to use this series

This suite of resources supports students and teachers following the Cambridge International AS & A Level Information Technology syllabus (9626). All of the books in the series work together to help students develop the necessary knowledge and critical skills required for this subject.

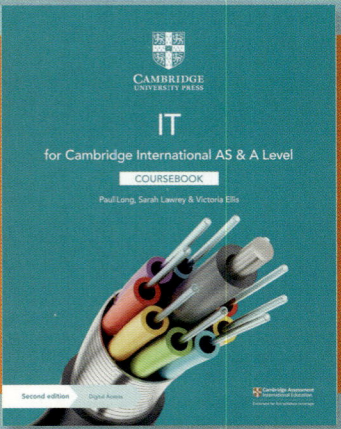

This coursebook provides 50% more practical activities than our previous edition and follows the same order structure as the Cambridge International AS & A Level Information Technology syllabus (9629). Each chapter includes questions to develop theoretical understanding or practical skills, and questions designed to encourage discussion. Exam-style questions for every topic help prepare students for their assessments.

The Teacher's Resource gives you everything you need to plan and deliver your lessons. It includes background knowledge at the start of each chapter, class activities with suggested timings, differentiation ideas, advice on common misconceptions, homework and assessment ideas.

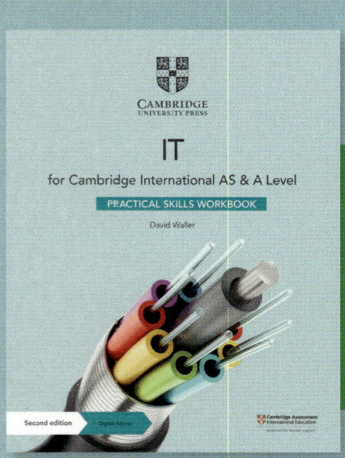

The Practical Skills Workbook contains worked examples and unique tasks to help learners practise core practical IT skills. With exercises increasing in challenge, it gives students further opportunities to undertake practice and refine their skills in the classroom or at home. It covers tasks for all of the practical chapters of the Coursebook that support many of the learning objectives required in the syllabus.

> How to use this book

When using this Workbook, please note that not all of the Learning Objectives from the Cambridge International AS & A Level Information Technology syllabus (9626) are covered. Those Learning Objectives covered in each unit are stated at the beginning of the unit. Throughout this book, you will notice lots of different features that will help your learning. These are explained below.

LEARNING INTENTIONS

A chapter outline appears at the start of every chapter to introduce the learning aims and help you navigate the content.

WORKED EXAMPLES

Wherever you need to know how to approach a skill, worked examples will show you how to do this.

PRACTICAL TASKS

This workbook provides activities for you to practise skills acquired from reading the Coursebook. Questions gradually progress in difficulty, to check and test your knowledge and understanding of the topic.

SUMMARY CHECKLIST

☐ The summary checklists are followed by 'I can' statements which match the Learning Intentions at the beginning of the chapter. You might find it helpful to tick the statements you feel confident with when you are revising. You should revisit any topics that you don't feel confident with.

SKILLS

Skills boxes appear at the start of tasks, and list the skills you will cover.

TIP

Facts and tips are given in these boxes.

KEY WORDS

Key vocabulary is highlighted in the text when it is first introduced. Definitions are then given in the margin, which explain the meanings of these words and phrases.

You will also find definitions of these words in the Glossary at the back of this book.

REFLECTION

Reflection activities ask you to look back on the topics covered in the chapter and test how well you understand these topics and encourage you to reflect on your learning.

> Introduction

The use of computers has freed users from having to carry out boring and repetitive tasks and also empowered them to carry out tasks for which they have no natural aptitude, such as drawing and painting.

The Cambridge International AS & A Level Information Technology syllabus (9626) contains many opportunities that allow you to carry out practical tasks that in the past had to be done by hand, from sending letters to lots of people to creating animations, carrying out calculations and storing data.

These applications even allow users with no drawing skills to demonstrate their creativity. Users can produce graphics and animations using pre-prepared images and clipart and also by designing and coding interactive programs.

The practical skills required to use these applications to their full potential are covered in this Workbook. The tasks are designed to cover many of the skills listed in the learning objectives of the syllabus but it does not cover all of them. You will find clearly stated learning intentions that outline these skills and develop in a progressive way with worked examples throughout.

We hope you will find the tasks interesting and exciting and that they encourage you to use the skills you will learn in your own projects.

Chapter 1

Algorithms and flowcharts

This chapter relates to Chapter 4 in the Coursebook.

The software used in this chapter is Word (Mac version).

LEARNING INTENTIONS

In this chapter you will learn how to:

- create an algorithm to solve a particular problem that demonstrates a decision-making process

- edit a given algorithm

- write an algorithm using pseudocode to solve a given problem

- edit a given flowchart

- draw a flowchart to solve a given problem.

Introduction

Algorithms are step-by-step instructions for solving a problem. In every area, algorithms are used to decide what action should be taken in a particular circumstance. As computers can consider all the possibilities far more quickly than a human brain, algorithms are becoming more important to the running of the world.

For example, in a game of chess, when each player has made three moves, there are over nine million possible moves available; after four moves there are over 288 billion possible moves. Computers have the ability to consider all these possible moves far more quickly than humans. That is why no chess grandmaster has beaten a top computer chess algorithm since 2005.

We use algorithms to carry out everyday tasks, often without thinking about them. For example, we use an algorithm or set of **procedures** to solve the complex problem of getting up and getting ready for school or college.

Algorithms can be displayed simply as written text or more formally using **flowcharts** or, when they are being designed to be used for a computer, as **pseudocode**. Pseudocode is a kind of structured English for describing algorithms. It is a generic, code-like language that can be easily translated into any programming language.

KEY WORDS

algorithm: a set of instructions or steps to be followed to achieve a certain outcome

procedure: a type of subroutine that does not return a value to the main program

flowchart: a set of symbols put together with commands that are followed to solve a problem

pseudocode: a language that is used to display an algorithm

WORKED EXAMPLE

As a security measure, users are required to enter their password before accessing a computer system. They are allowed three attempts before password entry is locked. Create an algorithm for doing this and display it as a flowchart.

The algorithm will contain:

- **Input**, e.g. the password. (Red)
- **Output**, e.g. informing the user if the password is recognised or not. (Blue)
- Decisions, e.g. is the password correct. (Green)
- **Process**, e.g. keeping count of the number of attempts that have been made. (Purple)
- A **variable**, e.g. to store the number of attempts that have been made. After each incorrect attempt it must be incremented by 1.
- Start and end symbols. (Black)

The algorithm will use conditional branching. There will be different branches when the password is recognised or not.

It will use a **loop**, which may need to be repeated up to a maximum of three times.

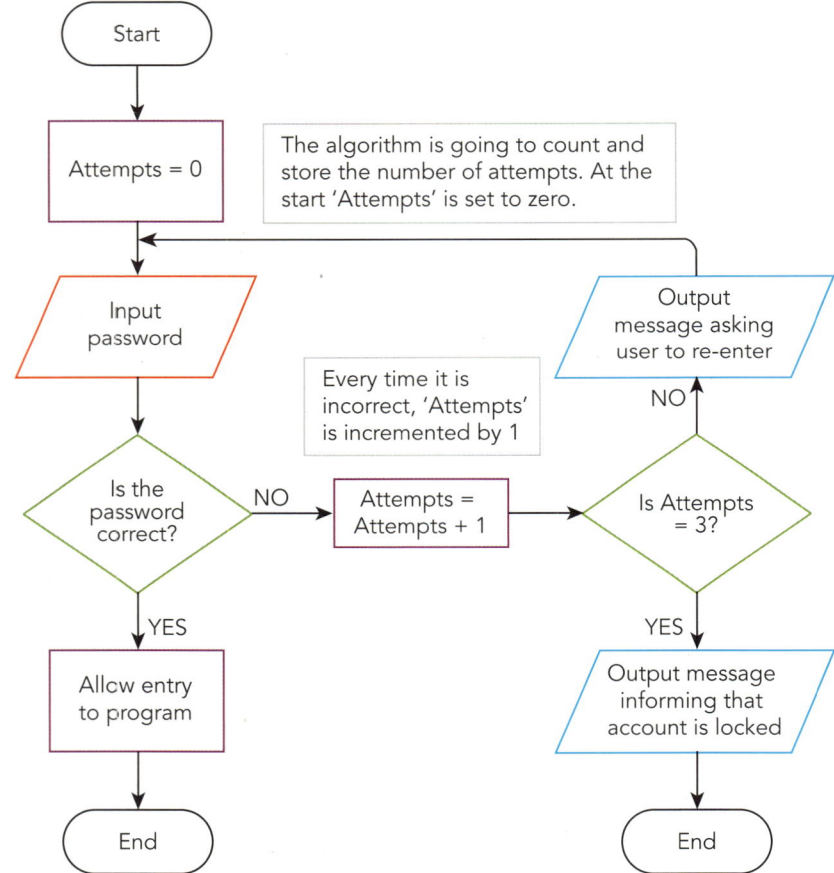

Figure 1.1: Flowchart of an algorithm that allows a user three attempts to enter a correct password.

Practical tasks

Task 1

SKILLS

This task will cover the following skills:

- drawing a flowchart with the following:

    - `input/output`

    - `decisions`

    - `start, stop and process boxes`

    - `connecting symbols / flow arrows`

- declaring variables and using arithmetic operators.

Create an algorithm that will ask a user to input the length and width of a rectangle and will then calculate the area.

Display your algorithm as a flowchart.

Task 2

SKILLS

This task will cover the following skills:

- `INPUT`

- `OUTPUT`

- arithmetic operators.

Display the algorithm from Task 1 in pseudocode.

Task 3

SKILLS

This task will cover the following skills:

- drawing a flowchart with the following:
  - input/output
  - decisions
  - start, stop and process boxes
  - connecting symbols / flow arrows
  - declaring variables and using arithmetic operators
- using a **count-controlled loop**.

KEY WORD

count-controlled loop: a loop where you know the number of times it will run

Draw a flowchart to display an algorithm to allow a user to enter ten numbers. The algorithm should then display the sum of the numbers.

Task 4

The flowchart in Figure 1.2 displays a partly completed algorithm for a game that simulates the throwing of three dice to find the player's score.

- If all three are equal, then the score is the total of the dice.
- If two are equal the score is equal to the sum of the two equal dice minus the third.
- If none are equal, then the score is zero.

1   Complete the flowchart of the algorithm.

2   Use the algorithm to calculate the scores from the following dice throws:

   a   3   6   3
   b   5   4   5

3   It is possible to obtain a negative score using the algorithm.

   State three dice numbers that would result in a negative score.

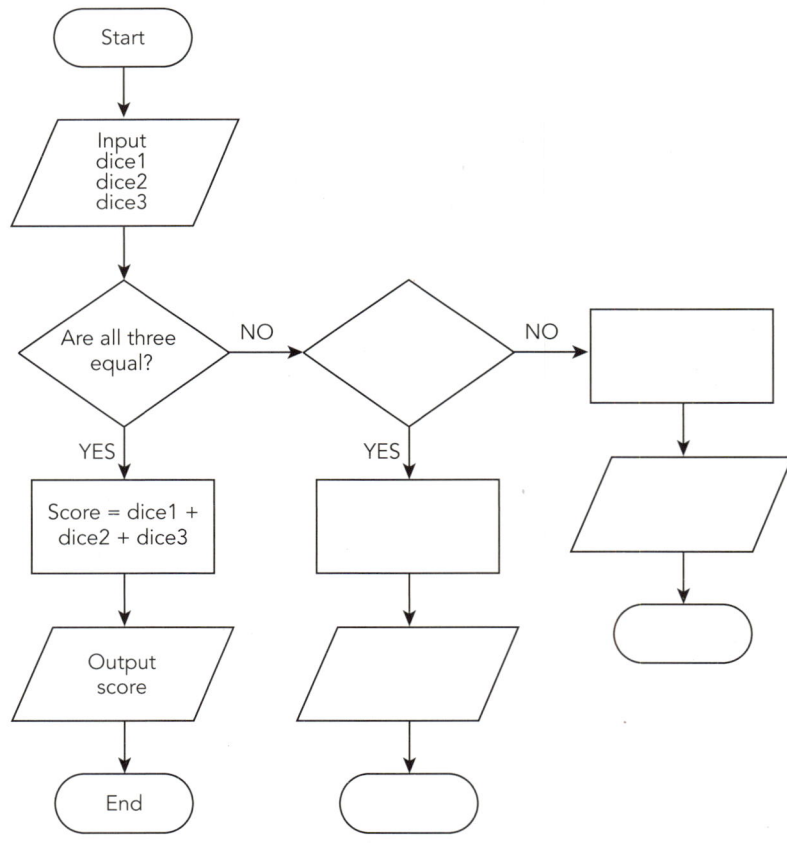

Figure 1.2: Flowchart of the algorithm to calculate the score.

Task 5

Create an algorithm to calculate the cost of sending a parcel using the following rules:

- If the weight of the parcel is 2 kg or under then the standard charge is $3.
- There is then a charge of $2 for each extra kilogram up to 10 kg.
- After 10 kg the charge per extra kilogram is $3.

Display the algorithm as

a   a flowchart.

b   pseudocode. Provide two versions of the algorithm. One should use a pre-condition loop and the other a post-condition loop.

Task 6

SKILLS
This task will cover this extra skill in addition to the ones you have already used: • **nested loops**.

KEY WORD

nested loops: one construct that is inside another construct

Create an algorithm that will output the times tables from 2 to 12.

It should state the table being output and then that number multiplied from 2 to 12.

Display your algorithm using pseudocode.

Task 7

SKILLS
This task will cover this extra skill in addition to the ones you have already used: • using CASE ... ENDCASE

In a multiple choice question, there are four possible answers, labelled A, B, C and D. To select their answer, users have to enter one of those letters. They will then be informed if they are correct or incorrect. There should also be a method to inform users if they have entered a character other than the four allowed.

For this question the correct answer is option C.

Create an algorithm to meet these requirements, using CASE...ENDCASE statements.

Task 8

A student handed in three homework assignments, which were each given a mark out of 10. All of the marks were different.

The student designed an algorithm to print out the highest mark.

Figure 1.3 is a flowchart of the algorithm, but some decision symbols are empty.

1   Complete the decision symbols and add 'YES' and 'NO' labels where required.

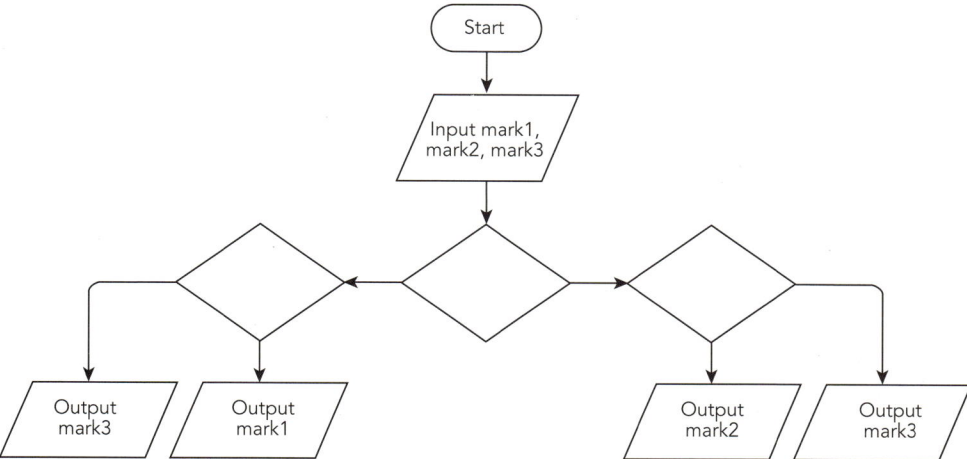

Figure 1.3: Flowchart of algorithm to print out the highest mark.

2   Display the algorithm using pseudocode.

Task 9

SKILLS
This task will cover these extra skills in addition to the ones you have already used: <ul><li>the use of procedures</li><li>the use of **parameters**.</li></ul>

KEY WORD
parameter: a piece of data that is sent to a subroutine

Design an algorithm that will allow a user to enter two numbers.

They should then be asked to enter either 'D' or 'M'.

If the user enters 'D', then a procedure should divide the first number by the second and output the result.

If the user enters 'M', then a procedure should multiply the two numbers together.

Create an algorithm and display it as pseudocode and as a flowchart.

Task 10

The following flowchart, Figure 1.4, displays an algorithm used by Holiday Theme Parks Limited to calculate the cost of customer entry, either individually or as part of a group.

1   Describe how the algorithm calculates the total amount that should be paid.

2   State two variables that are used in the algorithm.

3   In the flowchart, two of the constructs are labelled A and B. State the type of each construct.

4   The Smith family is visiting the park. The family consists of two children, one aged 8 and one aged 10, their two parents and their grandfather who is 65. Use the algorithm to calculate how much the family should pay for entry.

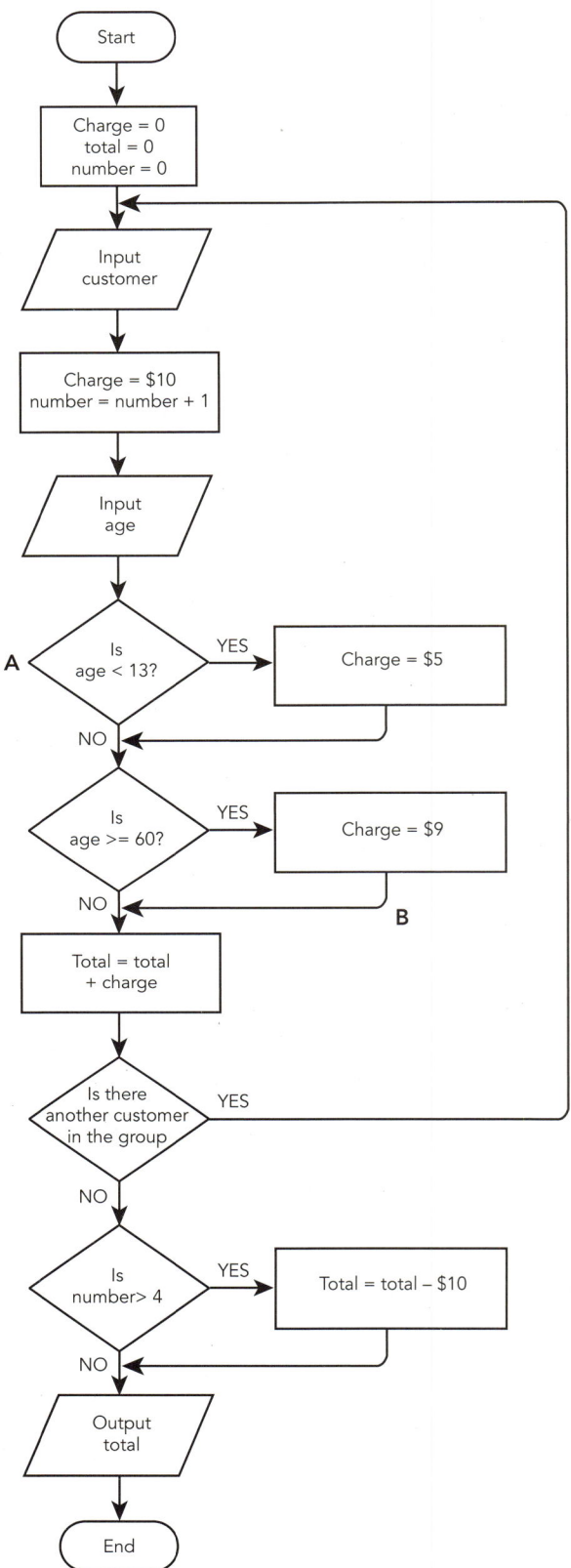

Figure 1.4: Flowchart to calculate the cost of entry.

Task 11

Part of an algorithm that a student has created to simulate the change given by a payment system is shown below. A user enters their payment and the algorithm determines the note and coins that should be returned.

```
1     charge ← RANDOMBETWEEN(1, 50)   // This creates
      a random number between 1 and 50. This line is
      correct.
2     payment ← 0
3     money ← 0
4     OUTPUT "Please enter payment"
5     INPUT money
6     payment ← payment + money
7     WHILE payment < charge DO
8         OUTPUT "The charge is ", charge, ". Please
          enter more money."
9     Input money
10    payment ← payment - money
11    ENDWHILE
12    change ← payment - charge
13    OUTPUT "Thank you. Change required is $, change
14    WHILE change >= 10 DO
15        OUTPUT "$10 note"
16        change  ← change - 10
17    ENDWHILE
18    WHILE change <= 5 DO
19        OUTPUT "$5 note"
20        change ← change - 5
21    ENDWHILE
22    WHILE change >= 2 DO
23        OUTPUT "$2 note"
24        change ← change +2
25    ENDWHILE
```

1 There are **four** errors in this algorithm. Some are logic errors and some are syntax errors. Identify the line numbers and state the correct version.

2 Explain the purpose of the loop between lines 7 and 11.

REFLECTION

- Before you used a formal flowchart or pseudocode on the computer, did you design the algorithm on paper?

- When you had written pseudocode or used a flowchart to display an algorithm, did you go through it using some test data to check it gave the correct result?

- Did you carry out online research, e.g. on loops or procedures?

SUMMARY CHECKLIST

- ☐ I can write an algorithm to solve a given problem.
- ☐ I can edit a given algorithm.
- ☐ I can use conditional branching.
- ☐ I can use looping.
- ☐ I can use nested loops.
- ☐ I can use procedures/subroutines.
- ☐ I can draw a flowchart to solve a given problem.
- ☐ I can edit a given flowchart.
- ☐ I can use input/output in a flowchart.
- ☐ I can use decisions in a flowchart.
- ☐ I can use start, stop and process boxes in a flowchart.
- ☐ I can use connecting symbols/flow arrows in a flowchart.
- ☐ I can use the following when writing pseudocode.
  - `INPUT/READ`
  - `WRITE/PRINT`
  - `IF... ELSE... ENDIF`
  - `WHILE... ENDWHILE`
  - `REPEAT...UNTIL`
  - `CASE...ENDCASE`
  - Comparison operators >, <, =
  - Arithmetic operators +, -, *, /
- ☐ I can identify errors in an algorithm/flowchart for a given scenario.

Chapter 2

Spreadsheets and modelling

This chapter relates to Chapters 8 and 9 in the Coursebook.

The software used in this chapter is Excel (Mac version).

LEARNING INTENTIONS

In this chapter you will learn how to:

- create structure:
  - create page/screen structures to meet the requirements of an audience and/or task specification/house style
  - create/edit spreadsheet structures
  - protect cells and their content
  - freeze panes and windows
- create formulas and use functions:
  - know and understand why absolute and relative referencing are used
  - use validation rules
  - format cells
- use a spreadsheet:
  - extract data
  - sort data
  - summarise and display data using pivot tables and pivot charts
  - import and export data
- automate operations with a spreadsheet:
  - create macros
- use graphs and charts:
  - create a graph or chart appropriate to a specific purpose
  - apply chart formatting
- use what-if analysis and goal seek.

Note: All the functions named in the syllabus are covered in this resource. However, some important additional functions are only covered in the Coursebook.

Introduction

The **spreadsheet** was the application that popularised the personal computer. VisiCalc, released in 1979 for the Apple II and in 1981 for the IBM PC, turned the personal computer from a hobby for computer enthusiasts into a business tool. It was a 'killer application' and, to a large extent, was responsible for Apple's success.

Today, spreadsheets are used in every company for organising, analysing and manipulating data, especially for financial operations but also for project planning and modelling real-life situations like technological developments in car design and performance and phenomena such as the weather.

KEY WORD

spreadsheet: software that can organise, analyse and manipulate data organised in a grid of rows and columns

WORKED EXAMPLE

BigCorp have sales people covering all areas of the world and use spreadsheet software to track their overall sales.

BigCorp_0.csv has been downloaded to suitable spreadsheet software and the 'Sales' figures have been formatted for currency.

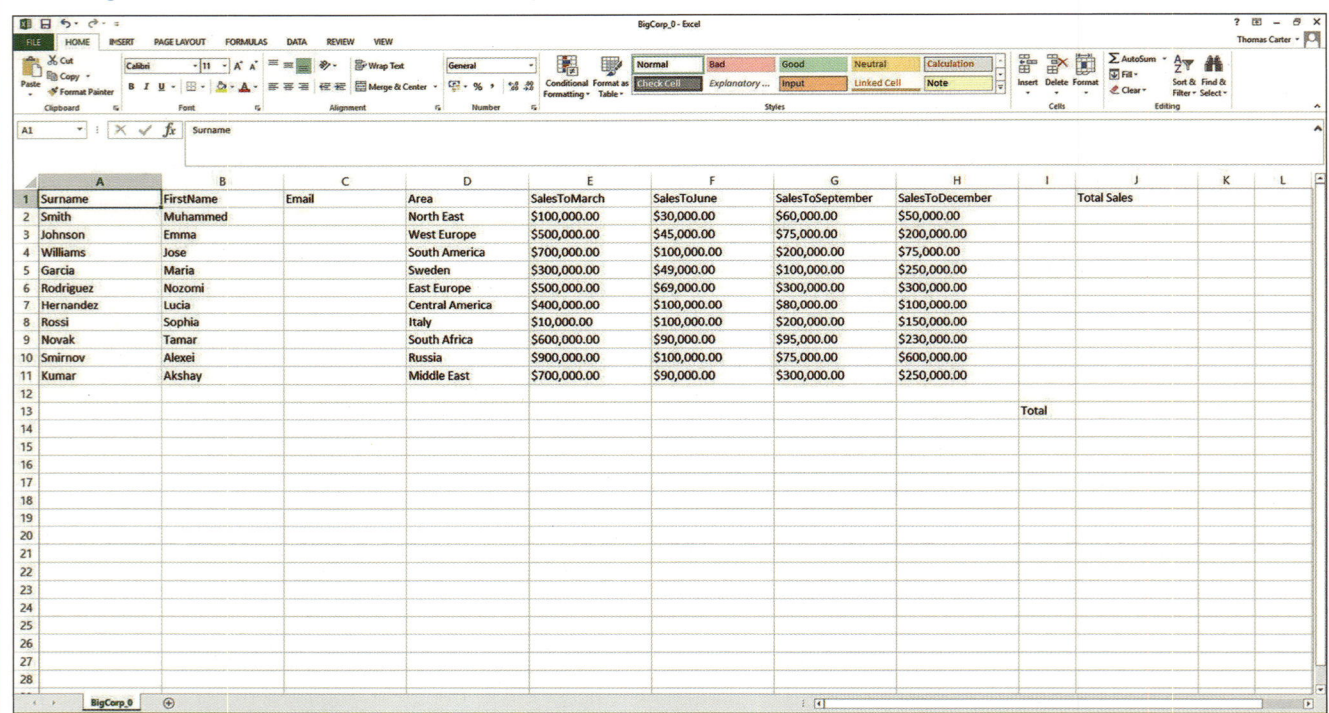

Figure 2.1: BigCorp's spreadsheet with figures formatted as currency.

You have been asked to develop the spreadsheet in the following ways.

1   Add the heading 'BigCorp Sales' at the top of the spreadsheet in a sans serif font with a size of 48 points. This heading should be centred across the first three columns.

CONTINUED

To do this a row has to be inserted above the existing row 1 and the text entered, resized and centred across the columns using 'merge and centre'.

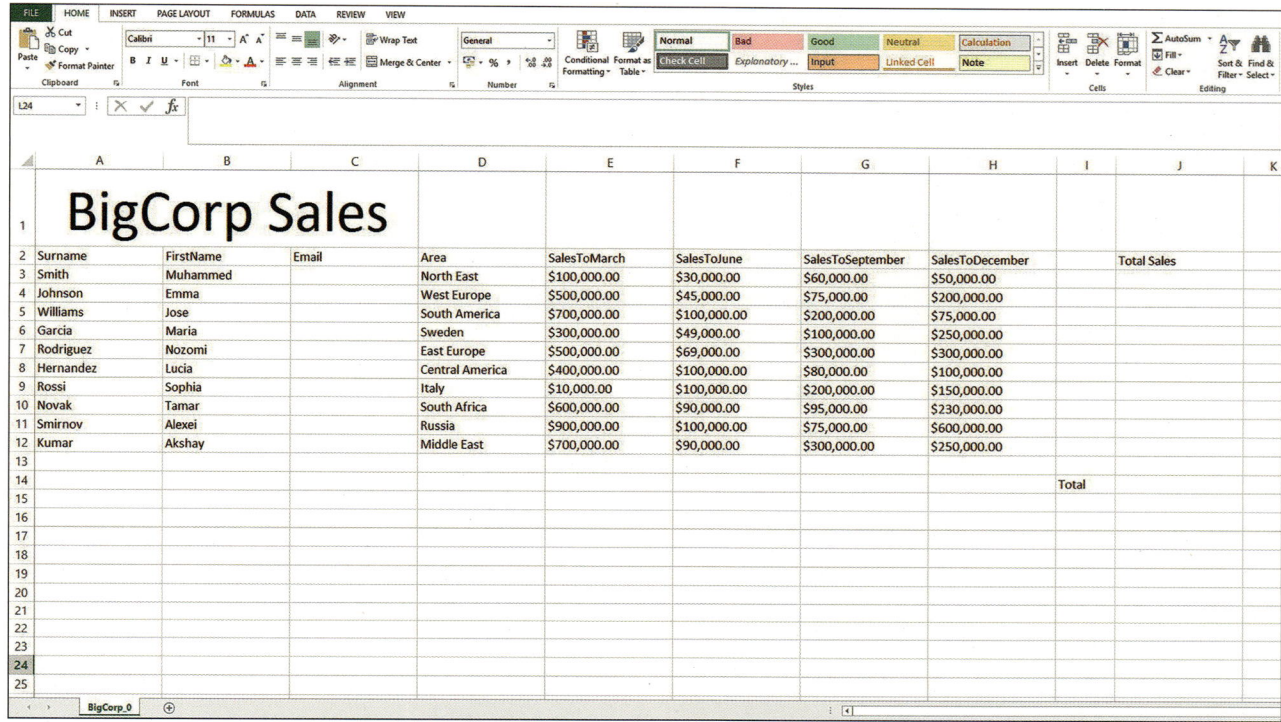

Figure 2.2: Spreadsheet with heading added.

2 a Use a **function** to create email addresses for the salespeople.

The email address should consist of the first letter of their first name, their second name and the text '@bigcorp.com'.

This can be done by using the CONCATENATE and LEFT functions.

```
=CONCATENATE(LEFT(B3,1),A3,"@bigcorp.com")
```

b Create **formulas** or use functions to calculate the total sales figures for each salesperson and the grand total.

This can be done by adding the function =SUM(E3:H3) in **cell** J3 and then copying it down to cell J12.

The function =SUM(J3:J12) can be used in cell J14.

KEY WORDS

function: a ready-made formula representing a complex calculation

formulas: mathematical calculations using +, −, × or ÷

cell: a single unit of a spreadsheet formed at the intersection of a column and a row where data can be positioned. Its reference (name/address) is based on its column letter and row number

CONTINUED

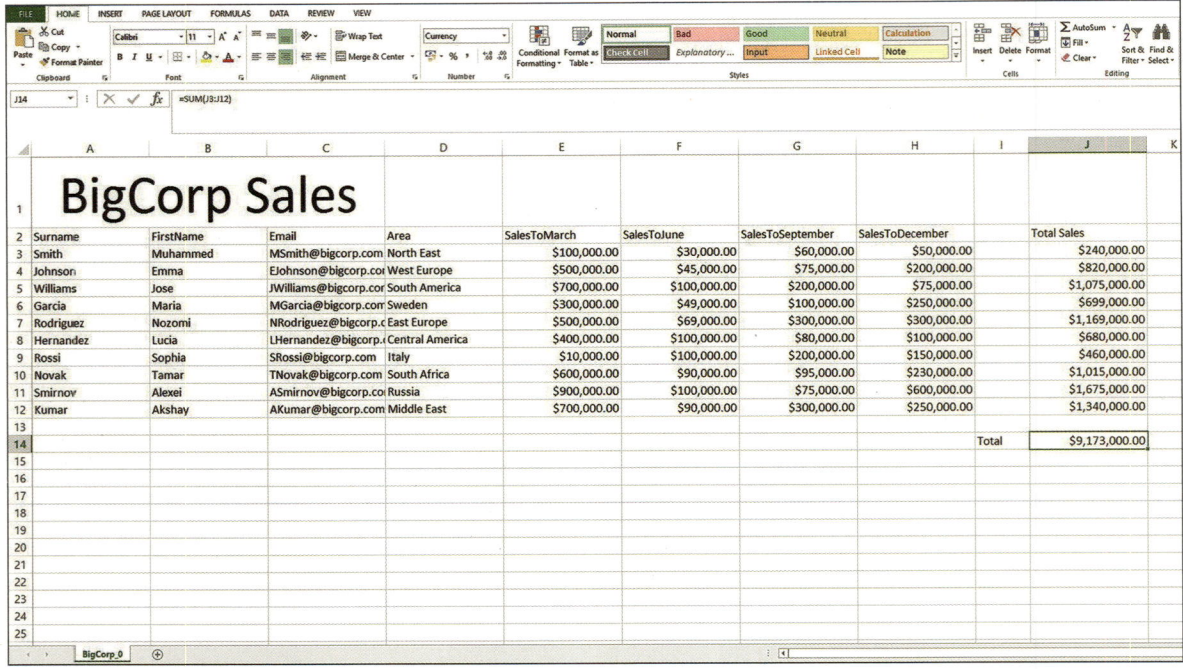

Figure 2.3: Spreadsheet with emails and total sales figures added.

3   In column K, show the percentage sales contribution of each salesperson.

This can be done by adding a suitable heading in cell **K2**. The cells **K3** to **K12** should be formatted for percentage, with one decimal place and the formula =J3/J14 placed in cell **K3**. This can then be copied down to cell **K12**.

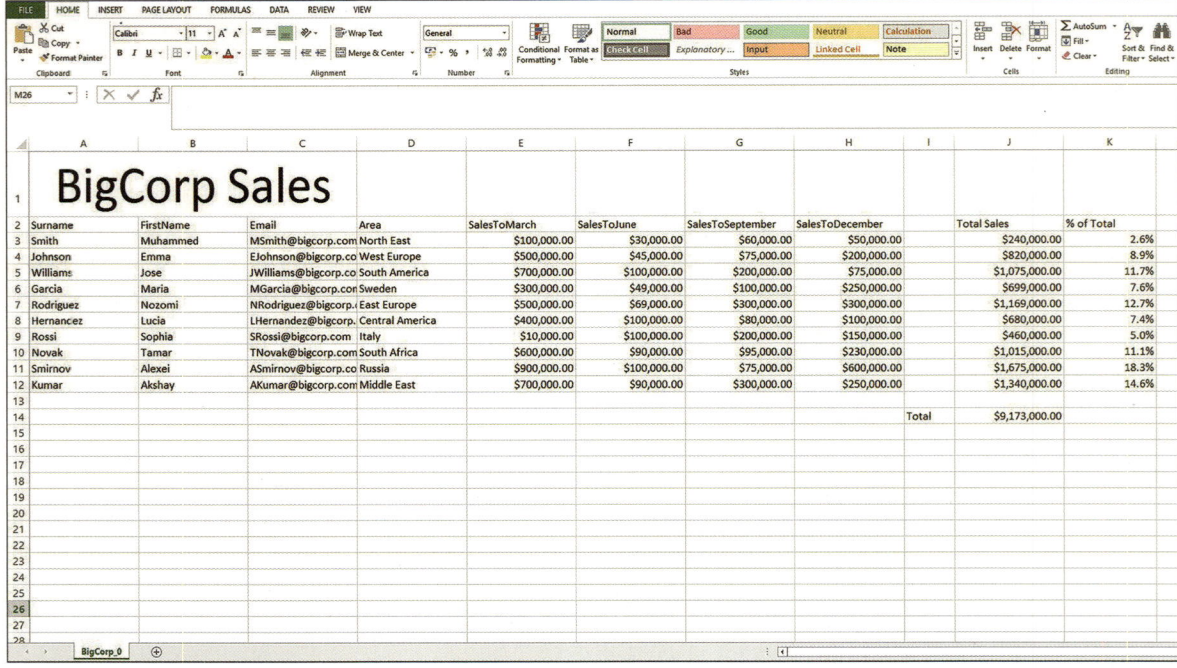

Figure 2.4: Spreadsheet with percentage sales contribution added.

CONTINUED

4   The cells containing the individual sales figures should be filled with red if they are less than the previous month and yellow if they are equal to or greater.

This is done using conditional formatting. Two rules for each cell can be set in one operation. If the **range** F3 to H12 is highlighted the rules can be set up.

The first formula for the yellow fill will be =F3 >= E3. The second formula for a red fill will be =F3 < E3. As a **relative** cell reference is needed the $ symbols, used for **absolute** referencing, have been removed. This formula will be applied to all cells relatively; for example, =G3 >= F3 and =G3 < F3.

KEY WORDS

range: a selection of cells

relative: a cell reference that changes when it is copied into other cells

absolute: a cell reference that does not change when it is copied into other cells, usually by placing a $ sign before the cell reference

---

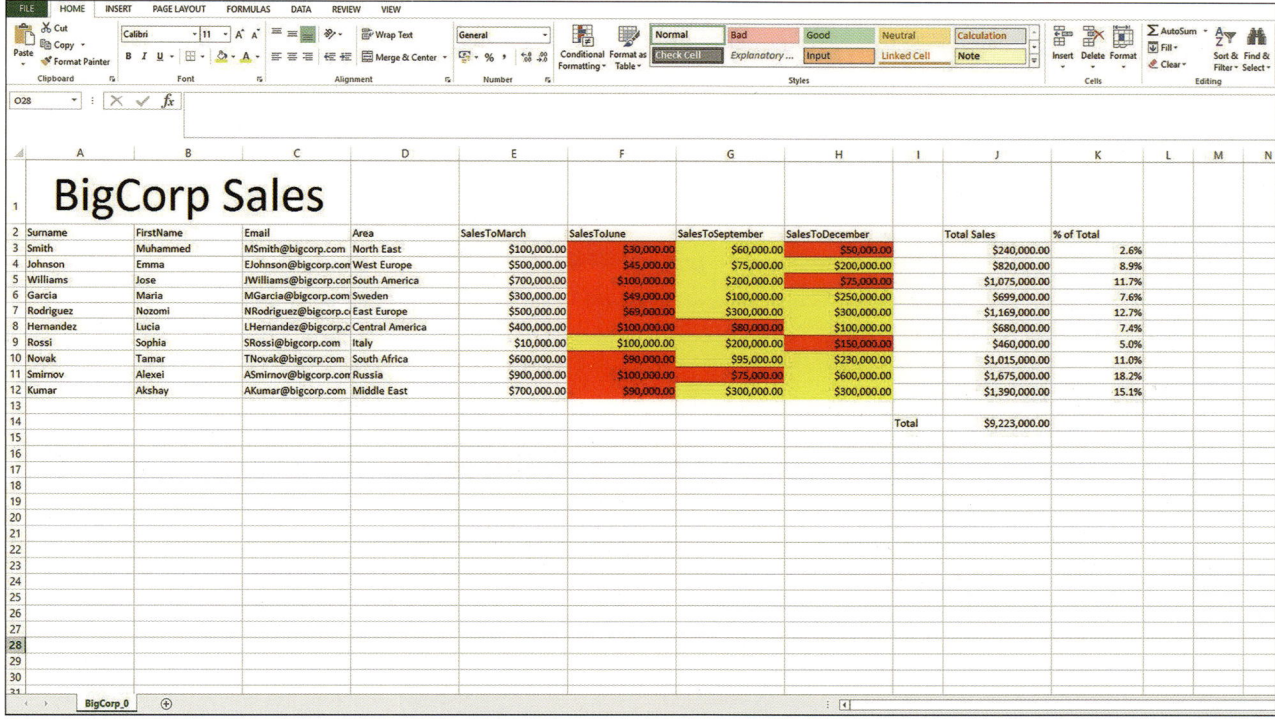

Figure 2.5: Spreadsheet with conditional formatting.

CONTINUED

5 In column L, a comment should be made for each salesperson using the following rules:

- If the percentage contribution is greater than or equal to 18%, the comment should be 'This is a superb result. You qualify for a 10% bonus.'

- If the percentage contribution is greater than or equal to 15%, the comment should be 'This is a very good result. You qualify for a 5% bonus.'

- If the percentage contribution is greater than or equal to 10%, the comment should be 'You have met your targets.'

- If the percentage contribution is less than 10%, the comment should be 'This is a poor result and must be improved next year.'

This can be done by using nested IF statements but you need to be careful of the following:

The IF function should not be IF(K3>=18, etc. because the numbers in cells K3 to K12 are all less than 1 but have been formatted as percentages. Therefore, the complete nested IF function should be:

=IF(K3*100 >= 18, "This is a superb result. You qualify for a 10% bonus.", IF(K3 * 100 >=15,"This is a very good result. You qualify for a 5% bonus.", IF(K3*100>=10, "You have met your targets.", "This is a poor result and must be improved next year.")))

Each value in cells K3 to K12 must be multiplied by 100.

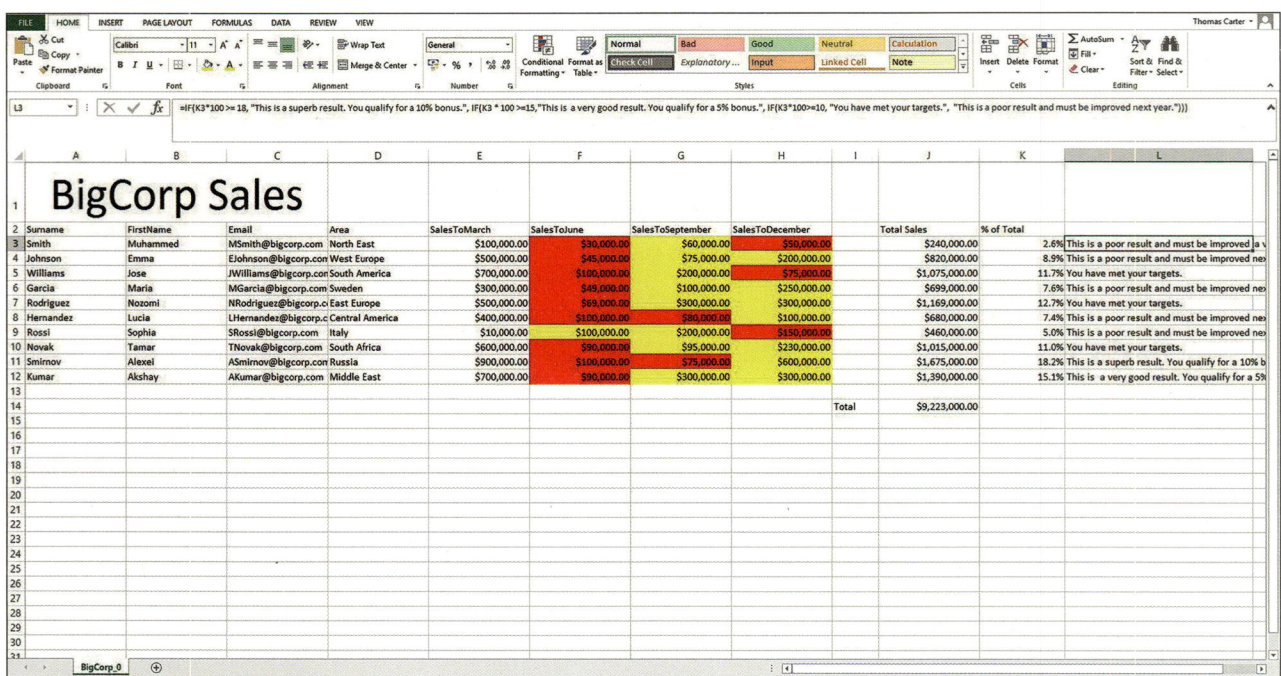

Figure 2.6: Spreadsheet with comments added.

CONTINUED

6 Finally create a two-dimensional bar chart showing the quarterly sales figures for each salesperson. The heading for the graph should be 'Sales per quarter for the year'.

This can be done by selecting the relevant data in columns A, E, F, G and H, selecting the required chart and adding the heading.

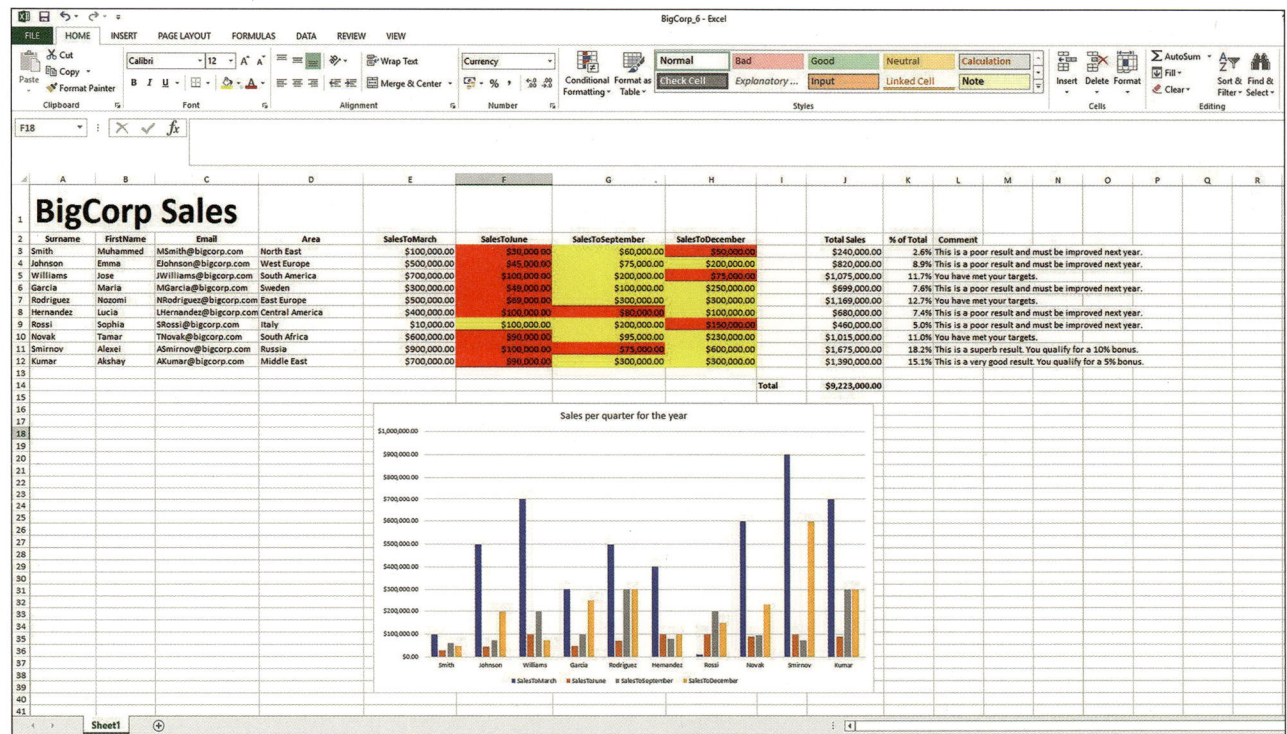

Figure 2.7: Spreadsheet with chart showing sales figures.

Practical tasks

Task 1

SKILLS

This task will cover the following skills:

- creating a spreadsheet
- creating/editing spreadsheet structures
- freezing panes and windows
- creating formulas and use functions
- using absolute and relative cell references
- formatting cells
- using conditional formatting.

This task will allow you to use a spreadsheet to solve the following problem.

A student has a Saturday job selling cups of tea and coffee. The tea is $1.20 per cup and the coffee is $1.90 per cup. He is supposed to keep a record of the number of cups of each he sells.

Unfortunately, he has been so busy that he has lost count, but he knows that he did not sell more than 100 of each.

He has collected $285.

Use a spreadsheet to calculate how many cups of tea and coffee he sold.

You can start with a blank spreadsheet or download **Task1_Spreadsheet.xls**.

To carry out this task:

- Use formulas to calculate how much it would cost for all combinations, i.e. from 1 to 100 of each.
- Some of the references will need to be relative and some absolute as the formulas will need to be copied to all cells.
- Freeze panes so that the tea and coffee column and row headings can always be seen.
- Use conditional formatting so that any cell containing $285 can be easily seen.

An example is shown in Figure 2.8:

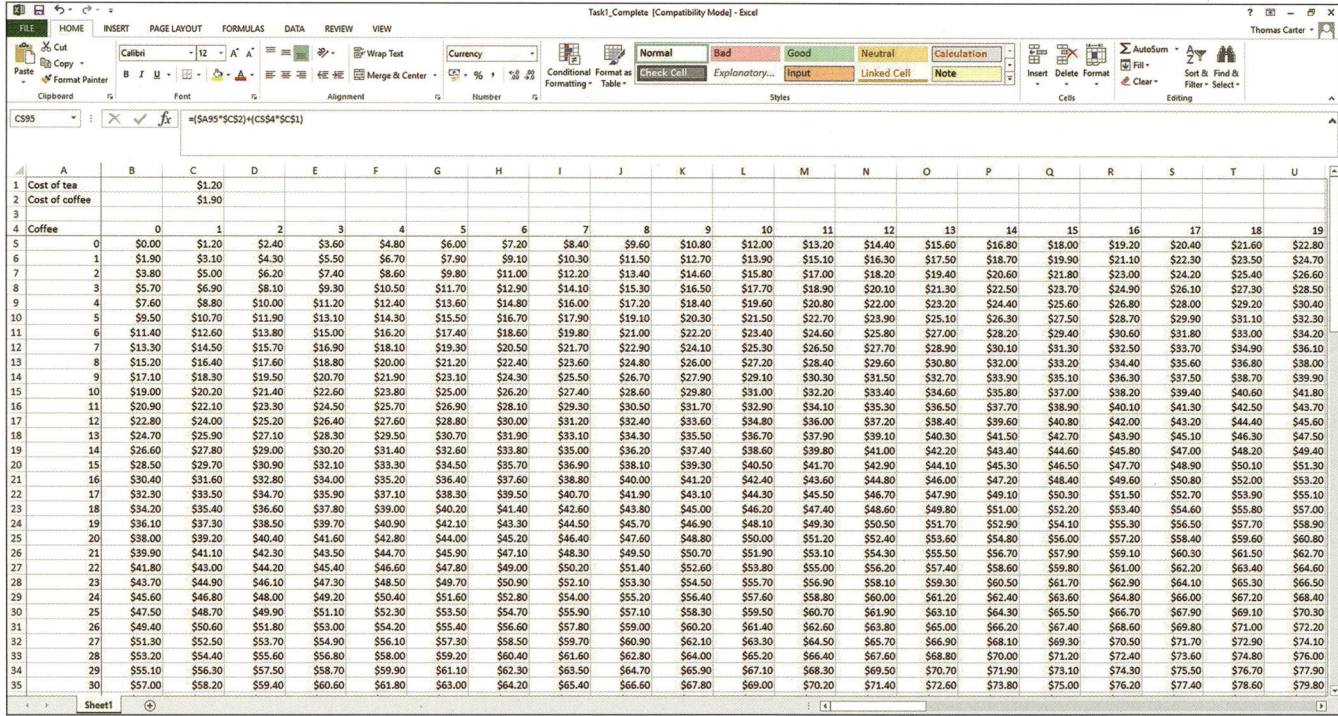

Figure 2.8: Spreadsheet to find the numbers of tea and coffee sold.

Task 2

KEY WORD

what-if analysis: experimenting with changing variables to see what would happen to the output if those variables changed

North Eastern College are staging a play on two consecutive evenings.

To help manage their ticket sales, costs, revenue and profit they have created a workbook containing four worksheets.

Download **Play_Q.xls** and examine the worksheets.

DataSheet contains data relating to costs, ticket, programme and refreshment prices and the percentage of the audience expected to purchase them.

Day1 and **Day2** contain blue-filled cells representing the front and rear seating. When a seat is booked, a 1 is placed in its cell and it turns red.

Totals contains cells to calculate the total costs, revenue and profit or loss.

1   Create a validation formula in all of the cells with a blue fill so that only a '1' can be entered.

2   Complete the workbook by adding formulas to the cells which are filled with yellow in all of the worksheets and then test the spreadsheet to ensure that all are working as expected.

3   On the Totals worksheet create a bar chart showing the revenues on each day, the total revenue and the profit.

Save your workbook as **Chapter02_Task2_Play_Q**.

4   The organisers are worried as tickets are not selling well.

They would like to use this model to investigate the following scenario:

- Only 24 of each of the front and rear seats are sold on each day. What percentage of the audience would they need to buy programmes to ensure that the plays didn't run at a loss?

Save your workbook as **Chapter02_Task2_Play_Q_Scenario**.

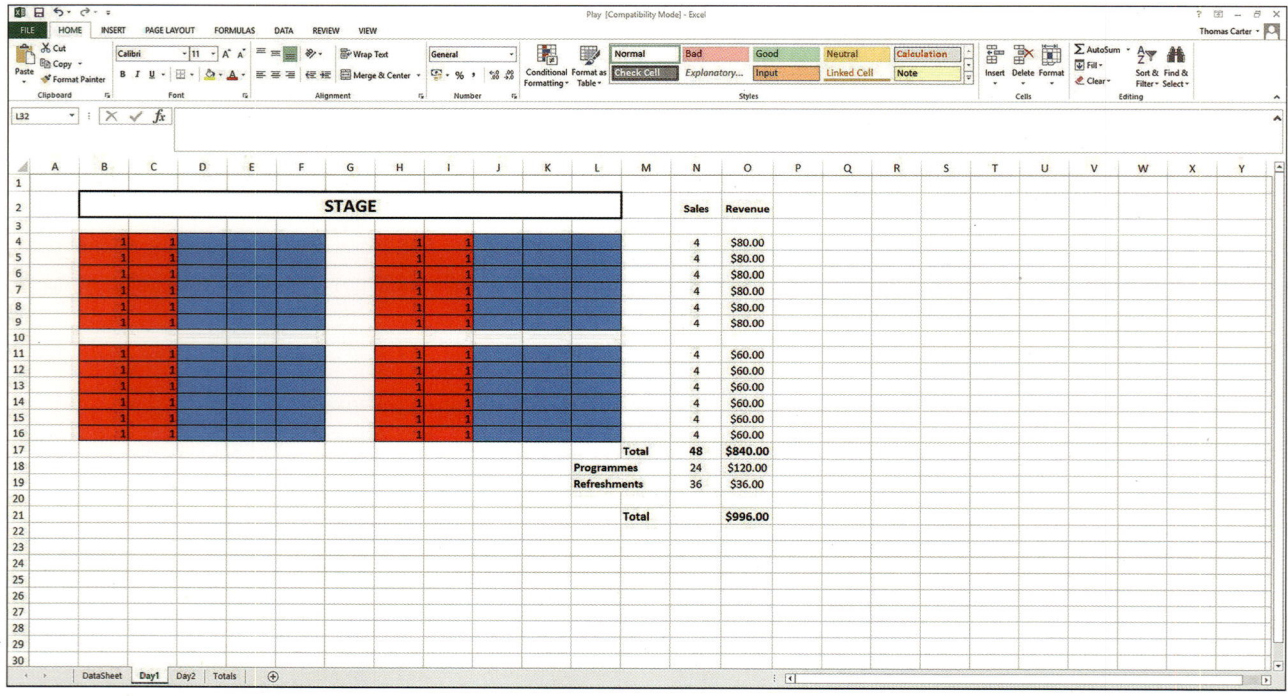

Figure 2.9: Spreadsheet showing 24 of each of front and rear seats sold.

Task 3

This task will allow you to use a spreadsheet to solve the following problem.

A student has a summer job helping to prevent and put out fires in a forest. He has a watchtower that is situated 300 metres from a water-storage canal, which he must use in case of fire.

One day, he spots a fire one kilometre from his position and 500 metres from the canal, as shown in Figure 2.10.

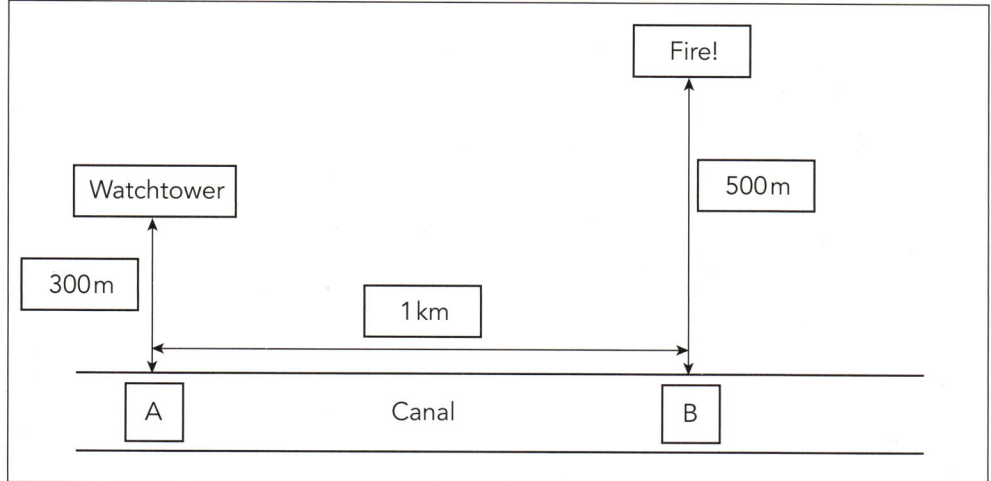

Figure 2.10: Plan showing watchtower, canal and location of the fire.

He could run straight down to the canal to collect the water, run along the canal bank and then run straight up to the fire. That route would be 1800 metres.

Or, he could run down to the canal at a point between the watchtower and the fire and then diagonally up to the fire.

What you have to do:

Use a spreadsheet to calculate the shortest route, to find both the distance of the route and the distance of the point on the canal from A where the water should be collected. Both distances should be given to the nearest metre.

A method you could use is to consider the route as two triangles and find the total of the hypotenuses.

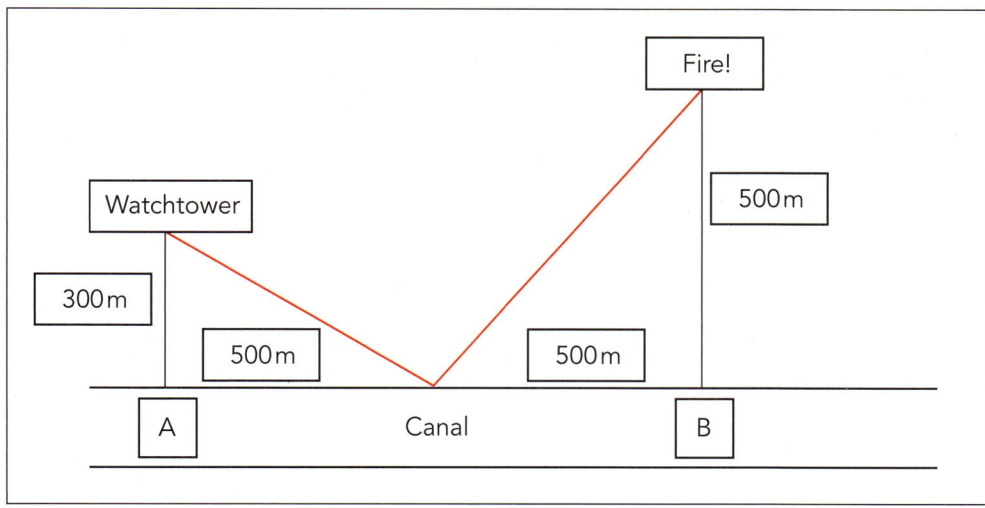

Figure 2.11: Possible route that the student could take.

Open **Task3.xls**.

1 Continue the numbering down column A up to 500.

2 Create a formula in cell B4, using relative and absolute references, that will calculate the total distance travelled if the student reaches the river at the number of metres from point A as shown in cell A4. Refer to cell D1 in your formula. Copy this formula down column B.

3 Use the minimum function in cell E5 to find the shortest distance.

4 Use INDEX/MATCH functions to find the distance from point A producing this shortest distance.

5 In cell E9 use the maximum function to find the longest distance with a suitable label in cell D9.

Save your completed spreadsheet as **Chapter2_Task3**.

Task 4

Open the file **Gym.csv** using suitable spreadsheet software.

1 Insert two rows above the column headings.

2 In row 1 enter the title 'The Gym' in a red, sans serif font, size 20 and centre it across columns A to G with a pale blue background.

3 In cell I1 enter the label 'Today' and in cell J1 place a function to show today's date.

4 In cell H3 enter the label 'Age (in years)' and in cell H4 enter a formula to calculate the age of the first member, using cell J1, in years. Copy this formula down column H for all of the members.

5 In cell I3 place the label 'Renewal Month' and in cell I4 enter a formula to show the month of the renewal date for the first member. It should be formatted to show the full month name. Copy this formula down column I for all of the members.

6 The gym would like you to create a table showing the sports as rows and the numbers of females and males who participate in those sports along with a grand total. They would also like a bar chart illustrating these data. Create these in a new sheet named 'Pivot'.

7 The gym would like another table with ages in years as rows and a count of the numbers of club members with those ages. They also require a chart showing this information.

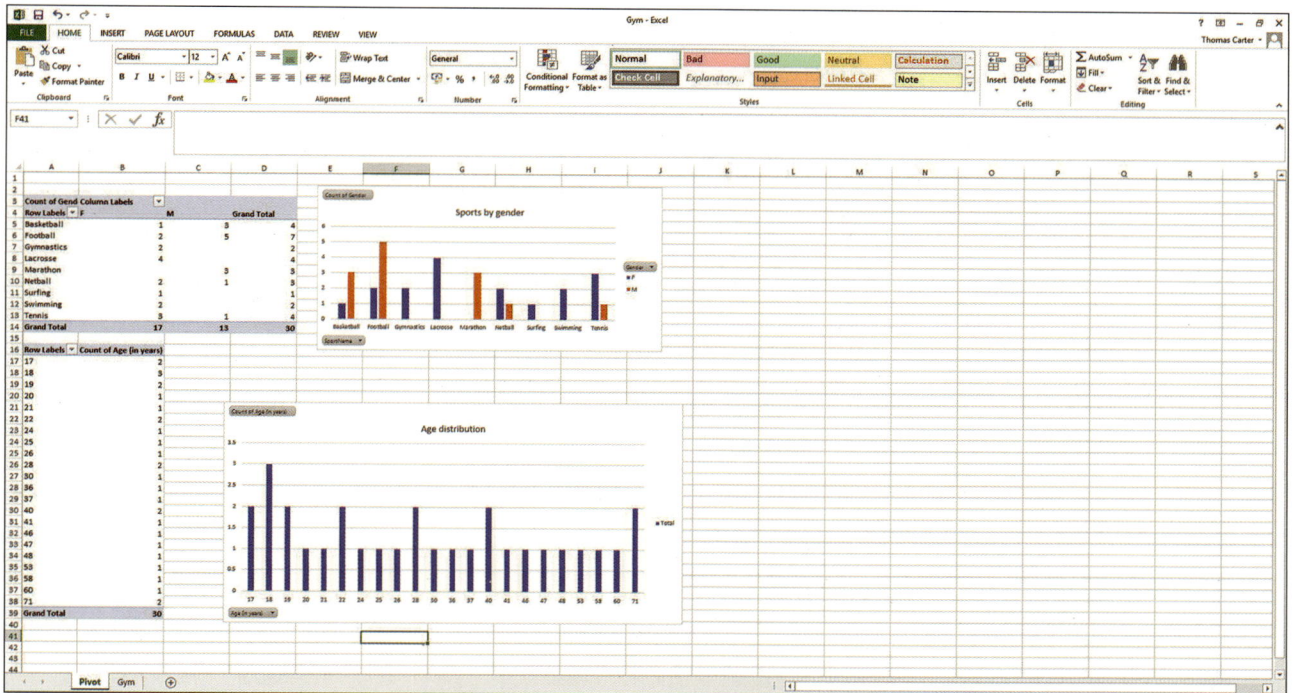

Figure 2.12: Spreadsheet with the two charts.

8 The members' main sports take place on different days and each one requires a supplementary payment.

The staff would like these to be shown on the main Gym worksheet.

You should do this using the HLOOKUP function. In a new worksheet create a table to hold information about the different sports and in columns J and K, use the HLOOKUP function to show the day and supplement for each member.

You can decide on the days and the supplements for each sport.

Save your spreadsheet as **Chapter2_Task4_Gym**.

	A	B	C	D	E	F	G	H	I	J	K	L
	AF42		×	✓	*fx*							
1					The Gym				Today	11/08/2020		
2												
3	MemberNumber	FirstName	Surname	Gender	DOB	RenewalDate	SportName	Age (in years)	Renewal Month	Day	Supplement	
4	272	Elisa	Saar	F	23/08/1993	31/01/2025	Lacrosse	26	January	Thursday	$15	
5	403	Agata	Lindberg	F	25/02/1961	15/03/2023	Swimming	59	March	Tuesday	$30	
6	579	Chiara	Moretti	F	03/04/1993	20/10/2022	Tennis	27	October	Wednesday	$30	
7	642	Enrico	Conti	M	08/03/1966	15/06/2023	Marathon	54	June	Wednesday	$10	
8	783	Bohai	Ho	M	23/06/1997	01/06/2025	Marathon	23	June	Wednesday	$10	
9	905	Minenhle	Okeke	F	31/03/1991	30/10/2022	Lacrosse	29	October	Thursday	$15	
10	1021	Laurent	Dubois	M	06/06/1978	08/10/2023	Basketball	42	October	Tuesday	$20	
11	1088	Yu Kiu	Lo	M	21/07/1990	15/04/2025	Football	30	April	Friday	$20	
12	1407	Honoka	Kimura	F	10/06/2000	01/01/2023	Netball	20	January	Thursday	$20	
13	1806	Matthew	Weber	M	17/11/1970	17/11/2021	Football	49	November	Friday	$20	
14	1925	Jamal	Khan	M	11/07/2001	12/09/2024	Tennis	19	September	Wednesday	$30	
15	1997	David	Hardwick	M	28/01/1948	23/07/2026	Football	72	July	Friday	$20	
16	2251	Rosa	Luxenburg	F	20/08/1977	01/05/2023	Gymnastics	42	May	Monday	$25	
17	2704	Kirill	Pavlov	M	15/03/1999	19/03/2023	Basketball	21	March	Tuesday	$20	
18	2846	Atallah	Zaman	M	02/10/1982	20/07/2022	Basketball	37	July	Tuesday	$20	
19	2881	Nicole	Moreau	F	31/07/1978	25/11/2025	Football	42	November	Friday	$20	
20	3005	Stephen	Jackson	M	12/05/1973	12/05/2024	Football	47	May	Friday	$20	
21	3122	Zakir	Bashir	M	20/08/1994	27/06/2023	Football	25	June	Friday	$20	
22	3206	Elina	Ricci	F	15/07/1981	28/10/2024	Surfing	39	October	Friday	$30	
23	3330	Jemima	Mustafa	F	29/10/2001	03/04/2026	Tennis	18	April	Wednesday	$30	
24	3385	Nikita	Jacobs	M	20/10/1996	20/10/2026	Netball	23	October	Thursday	$20	
25	3651	Valentina	Hoffmann	F	19/11/2000	11/11/2023	Tennis	19	November	Wednesday	$30	
26	3683	Bjorn	Karlsson	M	06/10/1971	27/06/2022	Marathon	48	June	Wednesday	$10	
27	4251	Catherine	Byrne	F	13/01/1948	13/01/2025	Netball	72	January	Thursday	$20	
28	4326	Sara	Mayer	F	13/06/1999	21/10/2024	Lacrosse	21	October	Thursday	$15	

Figure 2.13: Spreadsheet showing Day and Supplement for each user.

Task 5

SKILLS

This task will cover the skills you have used before and the following new ones:

- adding headers and footers
- carrying out a What-if analysis using **goal seek** and solver.

KEY WORD

goal seek: looking to see what a variable needs to change to for a goal in terms of output to be achieved

Sana has decided to use a spreadsheet to help her plan her finances for next year.

Open file **Finance.csv** using suitable spreadsheet software.

She has included her likely monthly income and costs, savings and total savings for the year.

1   Format the title 'Finance Planner' to a bold 20-point font and merge and centre it across columns A and B.

2   Add the words 'Finance Planner Spreadsheet' as a header. Then, insert a footer showing the page number and the total number of pages.

3   Add suitable formulas for the calculations in cells B7, B16, B18 and F18.

Format for currency with negative numbers shown in red.

Unfortunately, each month, Sana's costs are greater than her income (so she needs to ask her parents for some extra income) but she is desperate to save money towards a new car.

4   Use goal seek to find out how much she could spend on each of the following to save $500 over the year.

Note: each time do *not* accept the changes and return the values to the original ones.

a   Clothes.

b   Entertainment.

5   Sana does not want to spend less on clothes and entertainment. To what value would her allowance have to increase in order to save $500 at her current levels of spending?

Using goal seek, Sana can only change one value at a time, but she would like to see the solution if all values could be changed at the same time.

She is therefore going to use Solver, which is a spreadsheet add-in.

6   Use Solver, with the following constraints, to find a solution by changing the values in cells B10:B14.

B10 must be >= 50
B11 must be >= 20
B12 must be >= 10
B13 must be >= 75
B14 must be >= 50

Save your spreadsheet as **Chapter02_Task5_Finance**.

Task 6

<table>
<tr><td>SKILLS</td><td>KEY WORD</td></tr>
<tr><td>This task will cover the skills you have used before and the following new ones:

•   using VLOOKUP functions

•   using the IFERROR function

•   recording macros

•   writing macros

•   assigning macros to buttons.</td><td>macro: a set of instructions that can be completed all at once</td></tr>
</table>

Hassan is creating a spreadsheet model of a shop point of sales terminal.

When a user enters a product number it should show the product description and its price. It should also calculate a running total.

Open the file **Shop.csv** in a suitable spreadsheet program.

1   Cut and paste product table onto sheet two of the workbook.

2   On sheet one, enter the title 'The Mini-Market' in cell A1.

It should be in a red, sans serif font with a size of 36 points and should be centred across columns A to D.

3   In cell B4 enter the text 'Product Number', in cell C4 'Description' and in cell D4 'Price'. All should be in a bold font.

4   In cell C7 enter the text 'Total' and in cell D7 a formula to find the total of all of the cells above it in column D. This formula should function even if the formula in cell D7 is moved down by adding another row above it.

5   Enter formulas using the VLOOKUP function into cells C5 and D5 to display the description and price when a product number is entered into cell B5.

    When these formulas are first entered, they will show an error message as there is no number in cell B5. Adjust the formula so that this error message is not shown.

6   Whenever a product number is processed and if another one is needed, a new row will have to be added and the formulas in cells C5 and D5 will have to be copied down into it. Record a macro to do this and attach it to a button with the label 'Next'.

7   When all of the products have been processed, a receipt will need to be printed. Write a macro to show a print preview of sheet 1 and attach it to a button with the label 'Print'.

8   Test your spreadsheet by entering several items and check that everything functions and the total is calculated as expected.

Save your spreadsheet as **Chapter02_Task6_Shop**.

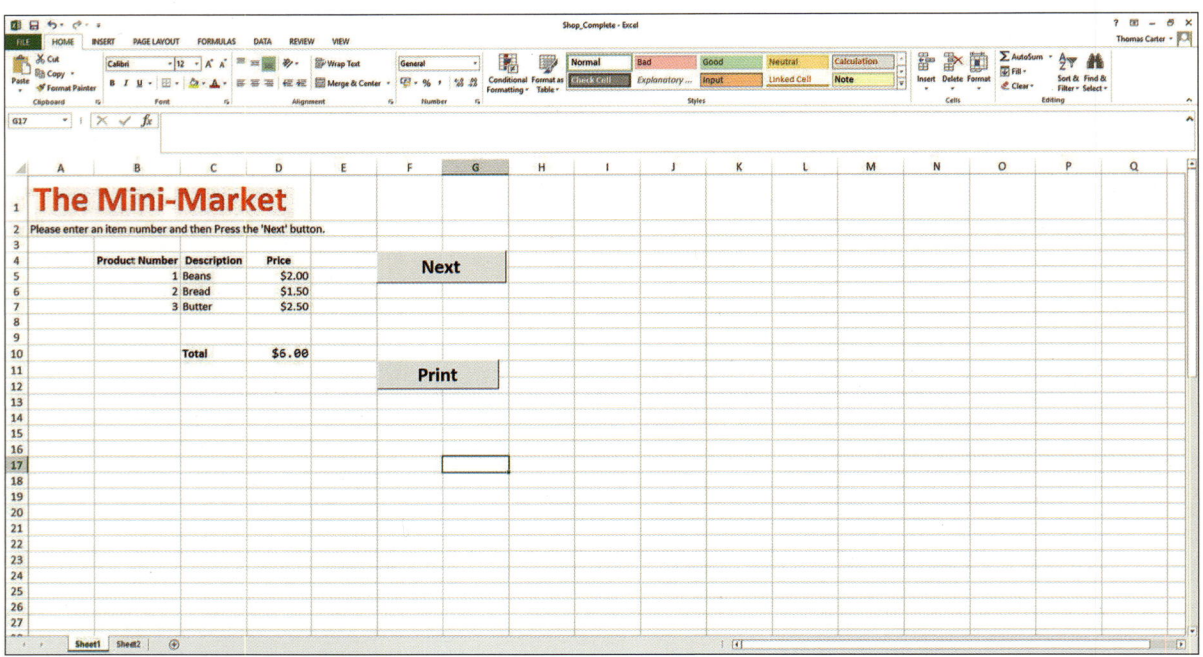

Figure 2.14: Example layout for the spreadsheet.

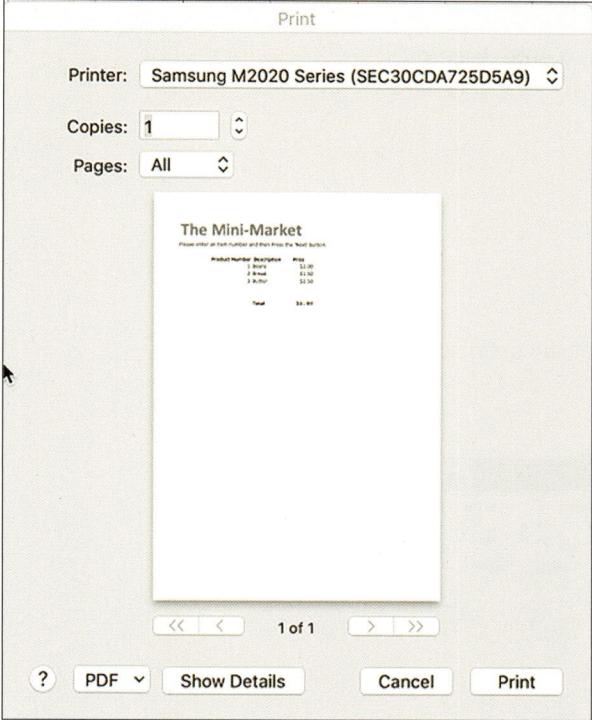

Figure 2.15: Example printed receipt.

9   Open the **Product_List** spreadsheet.

This shows the list when it was first delivered and unfortunately the prices have been inserted into the same column as the descriptions.

Your task is to use functions and formulae to extract the prices from the 'Description' column and copy them into the 'Price' column. The prices should then be deleted from the descriptions.

This is a relatively difficult task and so here is a clue – you should first find the position of the first digit.

Save your spreadsheet as **Chapter2_Task6_Product_List**.

Task 7

Your task is to create a pizza ordering system.

A pizza shop offers the following ingredients for buyers to design their own pizza:

Bases

Small $2
Medium $4
Large $6

All bases are supplied with tomato sauce and cheese.

Toppings

Olives, Chicken, Cajun chicken, Red peppers, Pineapple, Tuna

Mushrooms, Sweetcorn, Onion, Jalapeños, Chillies, Cheese

All toppings cost 90c and a customer can order as many as they want.

The customer should be asked if they will collect the pizza. There is a 10% discount if they collect it themselves.

Your task is to create a spreadsheet that will:

- Allow a customer to select their base.
- Allow customers to select their toppings.
- Allow them to select the option to collect it themselves.
- Show them the final price.
- Finally, it should have a button with an attached macro to reset the form so that nothing is selected.

Save your spreadsheet as **Chapter02_Task7_Pizza**.

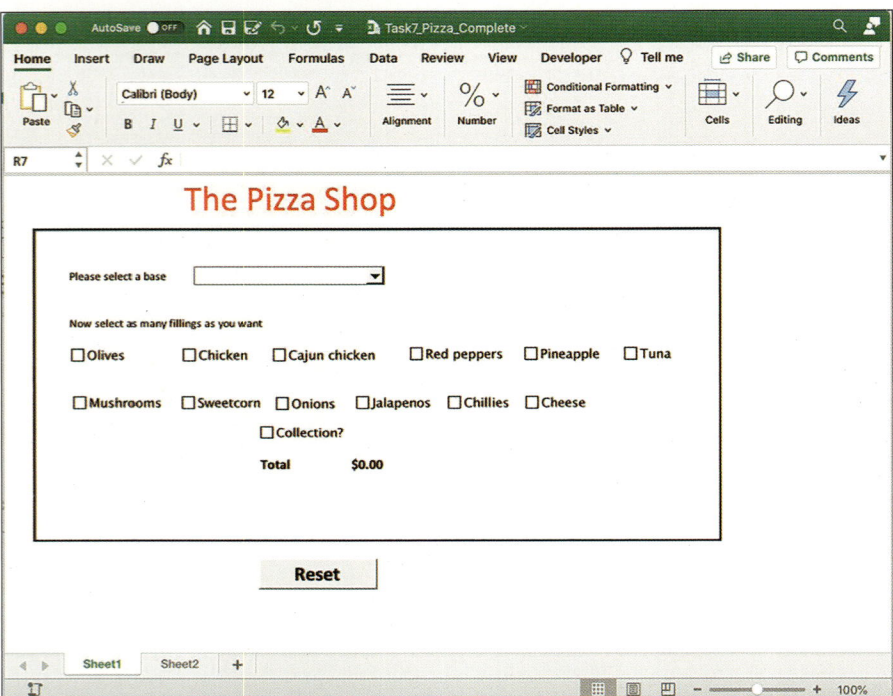

Figure 2.16: Example layout of the pizza ordering system.

Task 8

Open the file **Students.csv** using suitable spreadsheet software.

The file shows the results for 30 students in Maths, English and some of their option subjects.

1   Use conditional formatting to highlight all scores that are less than 50.

2   In cell O1 add the heading 'Average'.

In cell O2 use functions to calculate the average mark for that student and display the result as an integer.

The college would like to analyse the results by subject.

3   Starting in cell A35 create a list of all of the subjects and sort them into alphabetical order.

4   Use formulas to count the number of students studying each subject shown on the spreadsheet in column B.

5   Use formulas to sum the total scores achieved for each subject in column C.

6   Use formulas to calculate the average score for each subject and a function to round the number to one decimal place.

7   Create a line graph and a bar chart showing the average mark for each subject.

Save your spreadsheet as **Chapter02_Task8_Students**.

REFLECTION

- When you started to solve the problems did you think about them first and produce possible solutions or did you use the computer straight away?

- When you were creating these spreadsheets, did you test each formula and macro as you created them or wait until all of them were complete?

- Did you search online for help on using functions, combo boxes, recording macros, etc.?

SUMMARY CHECKLIST

- [ ] I can create structure.
- [ ] I can create page/screen structures to meet the requirements of an audience and/or task specification/house style.
- [ ] I can create/edit spreadsheet structures.
- [ ] I can freeze panes and windows.
- [ ] I can create formulas and use functions.
- [ ] I know and understand why absolute and relative referencing are used.
- [ ] I can use validation rules.
- [ ] I can format cells.
- [ ] I can extract data.
- [ ] I can sort data.
- [ ] I can summarise and display data using pivot tables and pivot charts.
- [ ] I can import and export data.
- [ ] I can create a graph or chart appropriate to a specific purpose.
- [ ] I can apply chart formatting.
- [ ] I can use what-if analysis and goal seek.

Chapter 3

Database and file concepts

This chapter relates to Chapter 10 in the Coursebook.

The software used in this chapter is Access.

LEARNING INTENTIONS

In this chapter you will learn how to:

- assign a data type and an appropriate field size to a field
- know and understand the three relationships: one-to-one, one-to-many and many-to-many
- create and use relationships
- create a relational database
- set keys
- validate and verify data entry
- perform searches
- use arithmetic operations
- design and create an appropriate data entry form including linked sub-forms
- design and create a switchboard menu within a database
- import and export data
- normalise a database to first normal form (1NF), second normal form (2NF) and third normal form (3NF)
- use static and dynamic parameters in a query
- use simple, complex, nested and summary queries (including cross-tab queries).

Introduction

People have been storing data for thousands of years. As soon as a writing system was invented in Sumer 5000 years ago, the rulers carried out a census and made lists of all the people and what they owned. This allowed them to start taxing them!

Database management was an obvious task for computers as they can search, sort and manipulate vast amounts of data far more quickly that a human.

KEY WORD

database: a structured method of storing data

Today, electronic and online databases are used in all areas of our lives to store data in a structured way, for example, information about students in a school or college, the items and customers in a shop and what the customers have bought, the aircraft, routes, timetables and passengers in an airline.

WORKED EXAMPLE

Details of some journeys provided by the taxi firm 'WeDriveAnywhere' are contained in the file **Taxi.csv**.

Taxi

TripID	Distance	NumberOfPassengers	StartTime
1	25	6	22:00:00
2	10	2	10:00:00
3	30	3	10:30:00
4	15	1	10:15:00
5	30	6	06:00:00
6	20	3	09:00:00
7	17	4	23:00:00
8	10	5	01:00:00
9	10	1	23:00:00
10	70	2	07:00:00

Figure 3.1: CSV file of 'WeDriveAnywhere' data.

1 Create a database **table** structure with suitable **fields** and import this data.

Add a screen print of the table in design view to the evidence document.

Fields will be needed to store this data.

Field Name	Data Type
TripID	AutoNumber
Distance	Number
NumberOfPassengers	Number
StartTime	Date/Time

Figure 3.2: Database table for the data.

KEY WORDS

table: a collection of related data, organised in rows and columns (for example, about people, places, objects or events)

field: (a common word for an attribute) a category of information about an entity stored in a database table, for example, product name, product number, ISBN code

CONTINUED

To calculate fares, the taxi firm have the following rules.

Between 8 a.m. and 8 p.m., the following rules apply:

- $3 for the first kilometre and $1 for every additional kilometre.

- If there are more than four passengers, there is a charge of $2 for each additional passenger.

Between 8 p.m. and 8 a.m., the basic charge is doubled.

2 Create fields in your table to calculate:

- the basic cost of the journey based on the number of kilometre

- the surcharge if there are more than four passengers

- the time surcharge if the time is between 8 p.m. and 8 a.m.

- the overall cost of the journey.

Insert screen prints to show the design of these fields.

The fields will have to be calculation fields based on these rules:

- The field for basic cost will have to calculate the fare based on the rule that it is $3 for the first kilometre and 1$ for every kilometre after that.

- The BasicCost field will require the following calculation using the IIf() function. It is similar to the IF function in spreadsheets.

General	Lookup
Expression	IIf([Distance]>0,3+([Distance]-1),0)
Result Type	Long Integer
Format	$#,##0.00;-$#,##0.00
Decimal Places	Auto
Caption	
Text Align	General

Figure 3.3: BasicCost field in design view.

The expression states that if the distance is greater than zero, the cost will be $3 plus the distance minus the first kilometre.

- The PassengerSurcharge field will add $2 for every passenger over four.

General	Lookup
Expression	2*IIf([NumberofPassengers]>4,([NumberofPassengers]-4),0)
Result Type	Long Integer
Format	$#,##0.00;-$#,##0.00
Decimal Places	Auto
Caption	
Text Align	General

Figure 3.4: PassengerSurcharge field in design view.

This function multiplies the number of passengers over four by two or inserts a zero if there are no more than four passengers.

CONTINUED

- The TimeSurcharge field will double the basic price if the time is between 20.00 and 08.00.

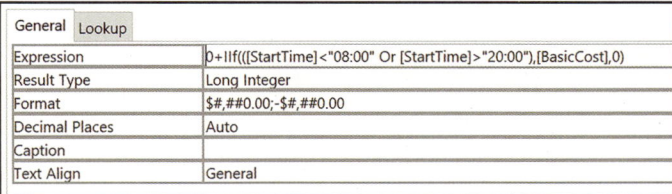

General	Lookup
Expression	0+IIf((([StartTime]<"08:00" Or [StartTime]>"20:00"),[BasicCost],0)
Result Type	Long Integer
Format	$#,##0.00;-$#,##0.00
Decimal Places	Auto
Caption	
Text Align	General

Figure 3.5: TimeSurcharge field in design view.

This expression inserts the basic price in this field if the time is less than 08:00 or greater than 20:00. Otherwise, it is set to zero.

- The TotalCost field calculates the total cost by summing these three fields.

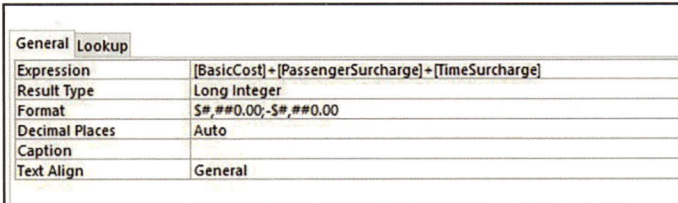

General	Lookup
Expression	[BasicCost]+[PassengerSurcharge]+[TimeSurcharge]
Result Type	Long Integer
Format	$#,##0.00;-$#,##0.00
Decimal Places	Auto
Caption	
Text Align	General

Figure 3.6: TotalCost field in design view.

All of these three fields are formatted for currency using the $ symbol.

3 Create a data entry form for this data.

Insert a screen print of the form.

The form can be created either in design view or by using the form wizard.
As we have not been asked to format it in a particular way, the wizard will produce a suitable form.

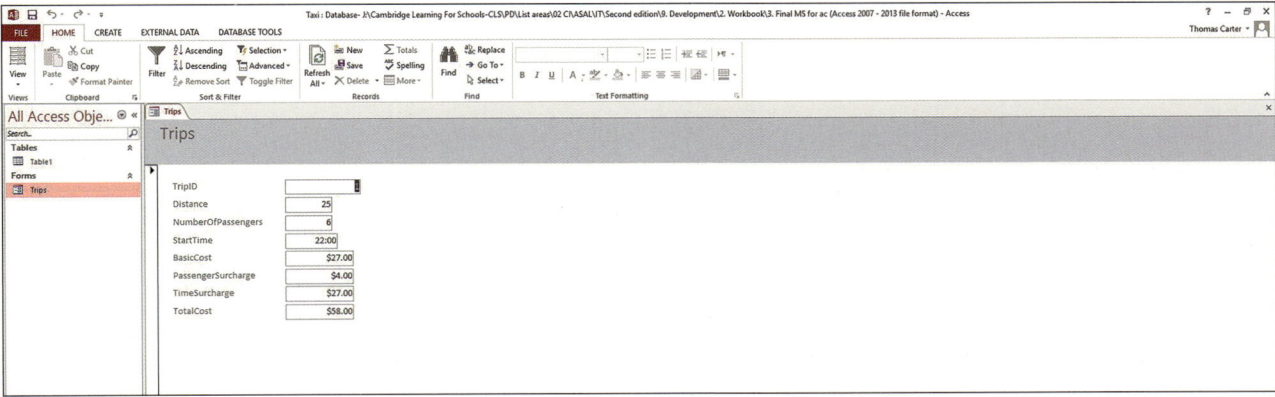

Figure 3.7: Data entry form for the table.

Practical tasks

Task 1

SKILLS
This task will cover the following skills: • assigning a data type and an appropriate field size to a field • setting keys • importing data • validating data.

GymFit is a gymnasium and sports centre offering facilities and sporting activities to its customers.

They would like to store all information about members, sports and teams in a computer database.

Their main priority is to store data about their members and they have asked you to create a database for this purpose.

The data they would like to store with specific examples are shown in Table 3.1:

Data item	Example
MemberNumber It must show the leading 0s	000272
FirstName	Elisa
Surname	Saar
Gender Should be either F or M	F
DOB It must ensure that members are at least 18 years old.	23/08/1993
eMail	ESaar@smile.com
ContactNumber It must have a fixed length and have a leading 0	01706580264
RenewalDate	31/01/2020

Table 3.1

1   Create a new database file named '**Gym**' and design a new table named '**tblMember**' to store data about this **entity** as shown in the table.

   a   Assign the correct data types, format and appropriate field sizes.

   b   For 'ContactNumber' apply an input mask that will require a user to enter 11 digits into this text field.

   c   For 'Gender' and 'DOB' apply validation rules with suitable validation text.

   d   Assign a suitable **primary key** field.

2   Import the data that is stored in the file **Members.csv**.

   To provide evidence of your work create a document named **Chapter03_Task1** and add to it the following:

   a   The fields in design view.

   b   The field properties for the 'Gender' field showing the validation rule and text.

   c   The field properties for the 'DOB' field showing the validation rule and text.

   d   The field properties of the 'ContactNumber' field showing the input mask.

   e   The error message when an incorrect entry is made in the 'Gender' field.

   f   The error message when an incorrect entry is made in the 'DOB' field.

   g   The input mask ready for data entry.

KEY WORDS

entity: a set of data about one thing (person, place, object or event)

primary key: a field that contains the unique identifier for a record

Task 2

SKILLS

This task will cover the following skills:

*   designing and creating an appropriate data entry form

*   using form controls

*   drop down menus.

1   GymFit have stipulated that they would like a data entry form to make it easier for their staff.

   The specifications for the form are:

   *   It should have a heading in red text on a blue background.

   *   All the fields and field names should be displayed.

   *   There should be no scroll bars or **record** selectors.

   *   There should be buttons for users to navigate the records and for adding and deleting members.

   Create the form and insert a screen print of it into a document named **Chapter03_Task2**.

KEY WORD

record: consists of all the fields about an individual instance of an entity in a database, e.g. all the details about one student

A sample is shown in Figure 3.8.

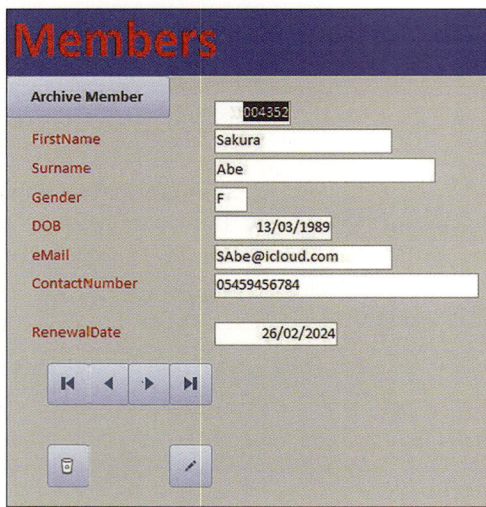

Figure 3.8: Data entry form for tblMember.

2   To make the form more user-friendly, GymFit would like you to include a drop-down menu (combo box) so that staff can select a member instead of having to scroll through them all.

You will need to add a macro to the control to move to the record of the member selected.

A sample is shown in Figure 3.9.

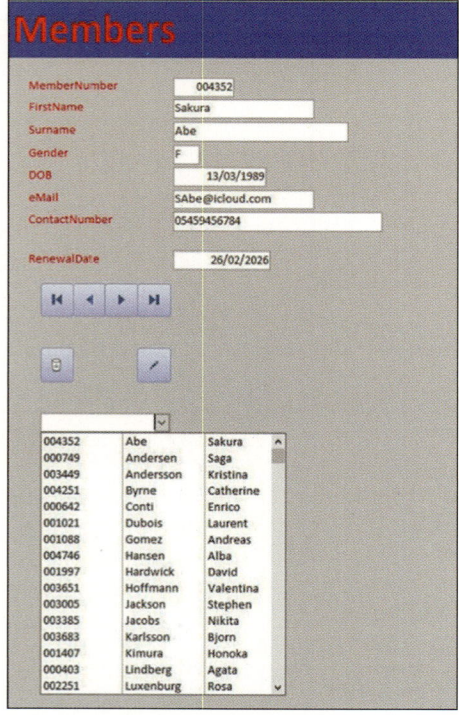

Figure 3.9: Data entry form with combo box to look up a member.

Insert a screen print of the form and open menu into the document **Chapter03_Task2**.

Task 3

GymFit would like the database to store information of the sports that the members take part in and the teams to which they belong.

A member of staff has changed the design of tblMembers to that shown in Figure 3.10.

Field Name	Data Type
MemberNumber	Number
FirstName	Short Text
Surname	Short Text
Gender	Short Text
DOB	Date/Time
eMail	Short Text
ContactNumber	Short Text
RenewalDate	Date/Time
Sport1	Short Text
Sport2	Short Text
Sport3	Short Text
Team1	Short Text
Team2	Short Text
Team3	Short Text

Figure 3.10: The tblMembers table showing changes made by a staff member.

1   Create a document named **Chapter03_Task3** and explain why this is not an acceptable design.

Following your advice, GymFit have suggested the following design.

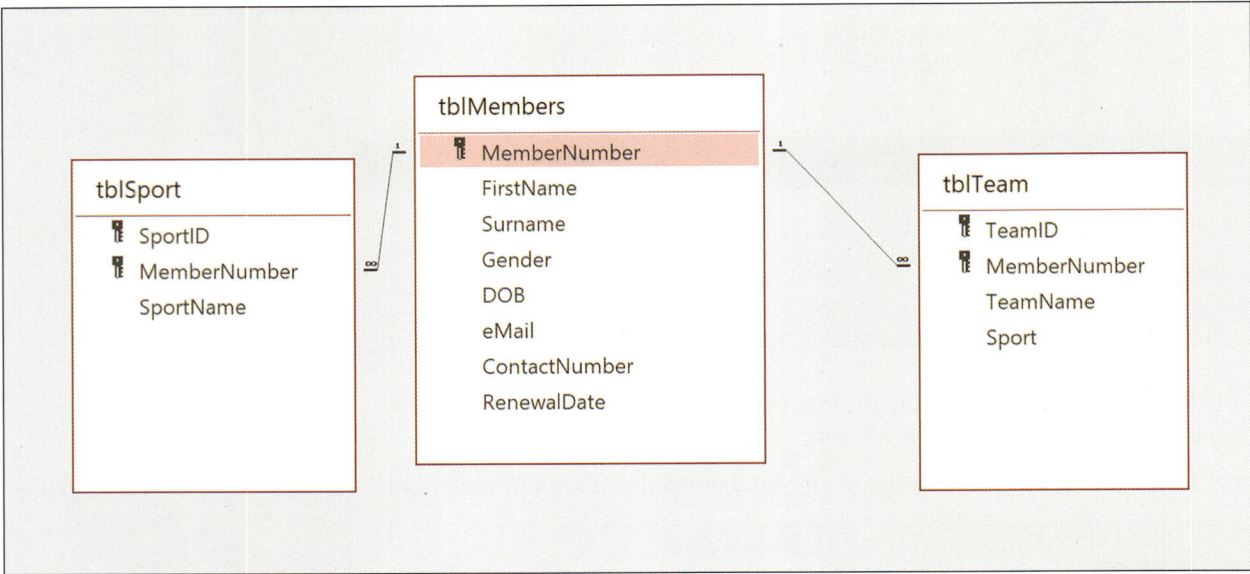

Figure 3.11: New database design.

2 In document **Chapter03_Task3**, explain why this design is not acceptable.

Again, following your advice, GymFit have changed the design to that shown in Figure 3.12.

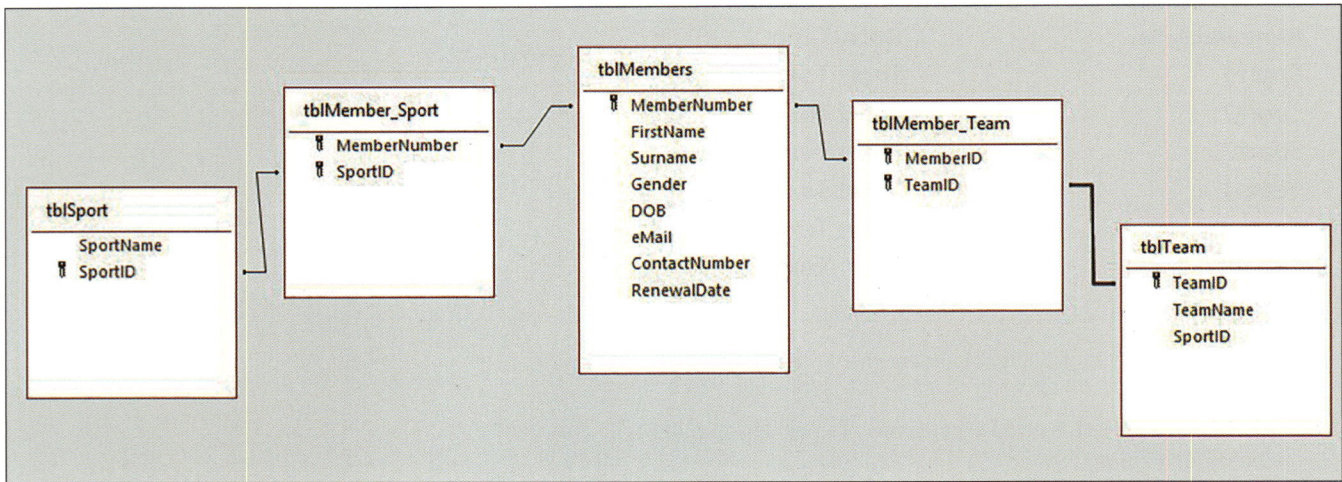

Figure 3.12: New database design showing new tables.

3 In document **Chapter03_Task3**, comment on this design.

Task 4

SKILLS

This task will cover the following skills:

- creating an entity relationship diagram
- assigning a data type and an appropriate field size to a field
- importing data
- designing and creating an appropriate data entry form.

It has been decided to use the design shown in Figure 3.12 in Task 3.

1   Create the tables tblSport, tblMember_Sport, tblTeam, tblMember_Team.

SportID should be a number that increments automatically when a new record is added.

TeamID should be formatted as text, with a length of 3 in upper case.

2   Import the data from file **tblSport.csv** into tblSport.

Import the data from file **tblTeam.csv** into tblTeam.

3   Import the data from **Member_Sport.csv** into tblMember_Sport.

Import data from file **Member_Team.csv** into tblMember_Team.

4   Set the **relationships** for these tables and stipulate that **referential integrity should be enforced**.

To provide evidence of your work create a document named **Chapter03_Task4** and add a screen print of this relationship diagram.

To make it easier for staff to enter and edit sport and team data, GymFit would like a form produced for each of tblSport and tblTeam.

5   Create forms for these tables.

- All the fields and field names should be displayed.
- There should be no scroll bars or record selectors.
- There should be buttons for users to navigate the records and for adding and deleting items.
- In the Team form there should be a combo box to add the SportID field.

To provide evidence of your work, add screen prints of these forms to your **Chapter03_Task4** document.

KEY WORD

relationship: the way in which two entities in two different tables are connected

Task 5

SKILLS
This task will cover the following skill: • creating linked subforms.

If the staff want to add records to the tblMember_Sport and tblMember_Team tables they will have to enter the correct MemberNumbers, SportIDs and TeamIDs.

This will be difficult and could easily lead to errors.

It has been decided that forms will be made from tblMember_Sport and tblMember_Team tables and these forms will be subforms of the frmMembers form so that the MemberNumbers will be automatically inserted.

1   Create the forms frmMember_Sport and frmMember_Team from the two tables.

In **both** forms, the user should be able to select a sport or a team from a combo box.

2   Position the two forms as subforms in frmMembers.

Remember to link the master and child fields correctly and to lock the MemberNumber controls on the sub-forms so that they cannot be edited.

An example is shown in Figure 3.13.

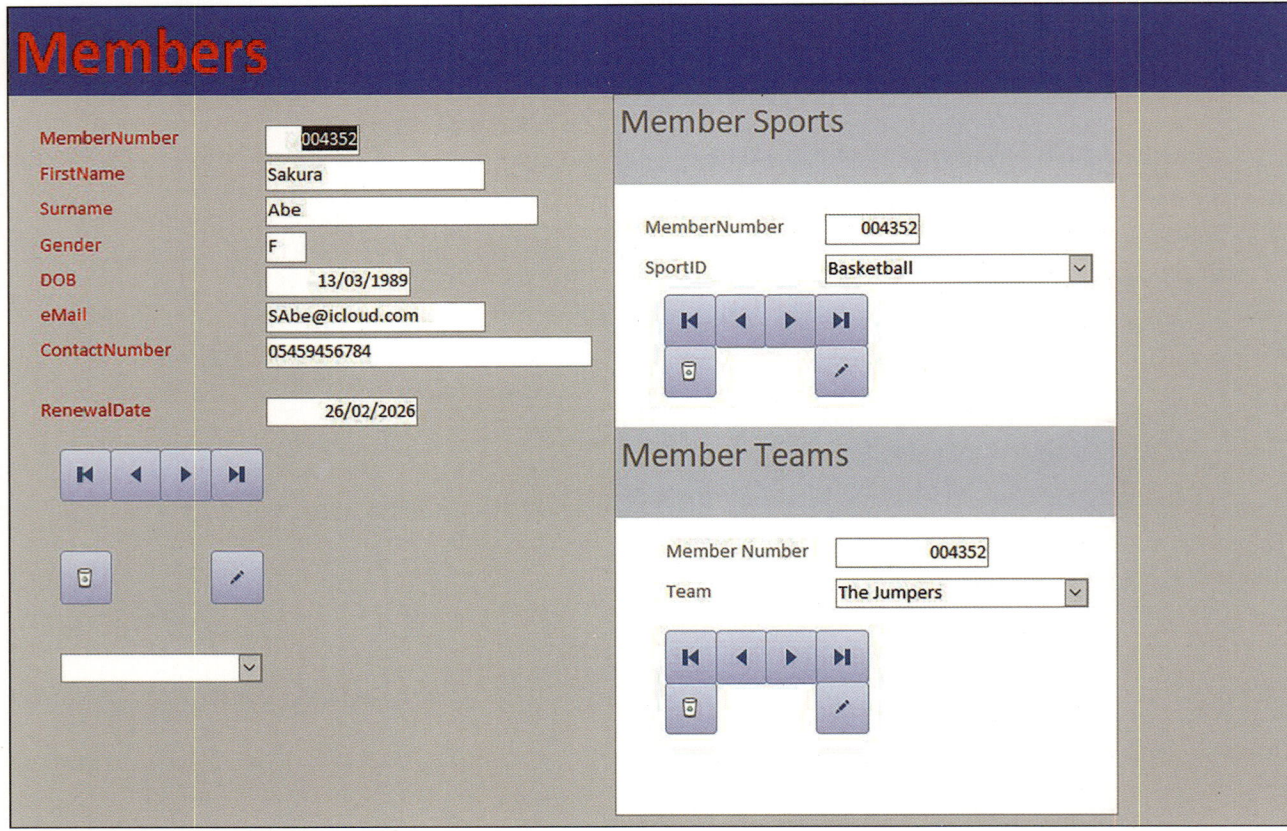

Figure 3.13: Data entry form with subforms.

To provide evidence of your work create a document named **Chapter03_Task5** and add to it a screen print of frmMembers with its two sub-forms showing details of the combo boxes.

Task 6

SKILLS

This task will cover the following skills:

- performing searches using single criteria and dynamic **parameters**

- designing and creating a database report.

KEY WORDS

parameter: the search criteria stipulated in a query

query: a request for data from a table or combination of tables

GymFit would like the users to be able to find members whose memberships will need to be renewed in the next month so that they can be send a reminder.

1   Create a **query** that will allow staff to enter two dates and return the details of all members who need to renew their membership between those dates.

   Name the query file as **qryRenewal**.

   Create a document named **Chapter03_Task6** and add to it a screen print of the query in design view showing the criteria.

2   Use qryRenewal to find all members who should renew their membership between 01/01/2023 and 30/06/2023.

   Add a screen print of your results to the **Chapter03_Task6** document.

   GymFit would like the results of the query to be displayed in a pop-up form for the staff.

3   Create a form, named frmRenewal, based on qryRenewal and place a button, with the label 'Renewal dates' on frmMembers to activate it using a macro.

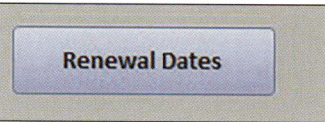

Figure 3.14: Button added to the form.

4   Open the form using the criteria between 01/01/2023 and 30/06/2023.

   Add a screen print of the open form to the **Chapter03_Task6** document.

5   GymFit would also like to send a reminder to members when their renewals are almost due.

   - Create a report, to send to each member, based on qryRenewal.

   - It should include their details and a message stating their renewal date.

   - It should be size A4 and have the gym name in red on a blue background.

   - Name this report **rptRenewal**.

   An example is shown in Figure 3.15.

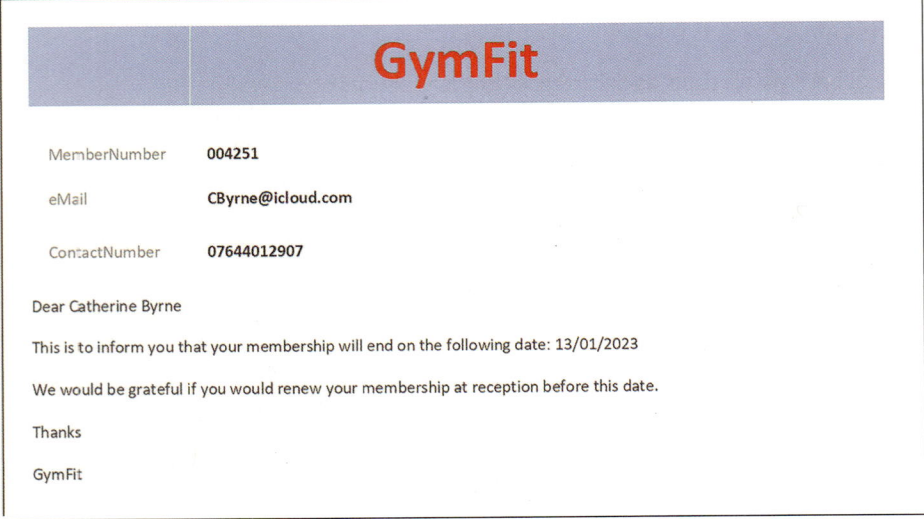

Figure 3.15: Report reminding users to renew their membership.

Add a screen print of the report in print preview view to the **Chapter03_Task6** document.

6   Add a button to frmMembers that will add one year to the renewal date of the member showing when they paid their subscription.

This button should run a macro that will increment the renewal date by one year.

Add a screen print of this function to **Chapter03_Task6** document.

> **TIP**
>
> You can use the DateAdd() function.

Task 7

> **SKILLS**
>
> This task will cover the following skills:
>
> - using a crosstab query
> - creating a grouped database report.

GymFit would like a query to show the number of males, the number of females and the total number of members taking part in each sport.

1   Use a query to summarise this data.

Create a document named **Chapter03_Task7** and add a screen print of your query in design view and a screen print showing the result of the query.

GymFit would also like a printed report analysing the teams.

2   The teams should be grouped by team name and also by gender, showing the totals for each gender and the grand total.

The fields required in the report are:

- TeamName
- SportName
- MemberNumber
- FirstName
- Surname
- DOB
- Gender

There should be a horizontal line between each team group.

The Team groups should be kept together when printing and there should be a page number showing the total number of pages on each printed page.

An example is shown in Figure 3.16.

Analysis of Teams

Boro **Football**

F

Membership Number	First Name	Surname	DOB	Gender
002881	Nicole	Moreau	31/07/1978	F
003449	Kristina	Andersson	10/03/1959	F
			Total:	2

M

Membership Number	First Name	Surname	DOB	Gender
003005	Stephen	Jackson	12/05/1973	M
001806	Matthew	Weber	17/11/1970	M
001997	David	Hardwick	28/01/1948	M
			Total:	3

Grand total: 5

Figure 3.16: Report showing grouping of data.

Export your report to PDF and add it to your evidence.

Add a screen print of your report in print preview, showing the complete first page.

Task 8

Staff at GymFit have raised issues about when members leave and then rejoin. They delete the members when they leave but then have to go to the trouble of entering all their details again when they rejoin.

GymFit would like you to do the following:

- When members leave, transfer all of their details to an archive table before they are deleted from the members table.
- Create a form that will show all the archived members and allow a user to rejoin the gym by adding them back to the member table and deleting them from the archive table.

To do this you will need to create:

- An archive table with the same fields as the member table.
- Update and delete queries.
- An archive form to view the archived members.

When your work is complete create a document named **Chapter03_Task8** and insert screen prints of the archive form when a member has been deleted and the member form when that member has rejoined, showing details of the member in each.

TIP

Records in the member table are also linked to records in other tables and you will have to preserve referential integrity.

Task 9

SKILLS

This task will cover the following skill:

- designing and creating a switchboard menu within a database.

Finally, GymFit would like you to create a switchboard form where a user can open the member, archive, sports and team forms.

Create a document named **Chapter03_Task9** and insert a screen print of your switchboard form.

Task 10

SalesCorp supply IT and electrical products and office furniture to businesses.

At present they store information about staff, products and sales in a spreadsheet.

Open file **SalesCorp.csv**. This contains some of the data.

SalesCorp have realised that it would be better to use a database to handle this information.

1   Create the required tables to set up a normalised database to store and process this information.

    You may add fields where required.

    Where data has to be added into a table from another table, Lookup combo boxes should be used.

    Create a document named **Chapter03_Task10**.

    Insert into the document annotated screen prints showing the tables in design view.

2   Insert a screen print of the relationship diagram of the tables, showing the join types.

3   Where possible import data relevant to the files into the tables from the spreadsheet.

    For one table, you may have to enter data manually.

4   Insert screen prints of the Lookup combo boxes in use in your table where sales are recorded.

SalesCorp would like the database to produce a report on the salespeople and how many items they have sold.

It should:

- show all of the salespeople with the items they have sold and the price of each

- calculate the total number of items sold and the total revenue for each salesperson

- have 'SalesCorp' inserted as a heading. It should also include a footer showing the page number and the total number of pages

- not split the data for a salesperson between pages.

5 Create the report and insert a screen print of the first page in print preview.

REFLECTION

- When you create a database, do you plan the structure, tables, fields, relationships, etc. before you start or use the computer and just let it develop?

- When you look at the tables in a database, can you easily see if it is in first, second and third normal form?

- When you plan the structure of a database, do you always ensure that there is no data redundancy?

SUMMARY CHECKLIST

- [ ] I can assign a data type and an appropriate field size to a field.
- [ ] I know and understand the three relationships: one-to-one, one-to-many and many-to-many.
- [ ] I can create and use relationships.
- [ ] I can create a relational database.
- [ ] I can set keys.
- [ ] I can validate and verify data entry.
- [ ] I can perform searches.
- [ ] I can use arithmetic operations.
- [ ] I can design and create an appropriate data entry form including linked sub-forms.
- [ ] I can design and create a switchboard, menu within a database.
- [ ] I can import and export data.
- [ ] I can normalise a database to first normal form (1NF), second normal form (2NF) and third normal form (3NF).
- [ ] I know and understand the use of static and dynamic parameters in a query.
- [ ] I know and understand when simple, complex, nested and summary queries (including cross-tab queries) should be used.

> Chapter 4

Sound and video editing

This chapter relates to Chapter 11 in the Coursebook.

The software used in this chapter is:

- Audacity (free software) for audio editing
- iMovie on the Mac for video editing.

LEARNING INTENTIONS

In this chapter you will learn how to:

- edit a video clip to meet the requirements of its intended application and audience
- know and understand why typical features found in video editing software are used
- edit a sound clip to meet the requirements of its intended application and audience
- know and understand how and why typical features found in sound editing software are used.

Introduction

Sound and video recording and editing have become very popular pastimes with the widespread use of personal computers and portable digital devices. Where once a dedicated studio was required, filled with expensive equipment, now anyone can record and edit on a smartphone using sophisticated and, often, free apps and share their creations on social media.

Everyone can be a film director or music producer.

> **WORKED EXAMPLE**

Open and examine the file **Cascade.mp4**. The clip is of a water feature.

1 Split the clip at 11 seconds.

Figure 4.1: Cascade clip.

2 Remove the video between 11 and 14 seconds.

This can be done by splitting the video again at 14 seconds and then deleting the clip between 11 and 14 seconds.

Figure 4.2: Removing video between 11 and 14 seconds.

3 Remove all the video between 22 seconds and the end of the video.

This can again be done by splitting and deleting. The video should now be 22 seconds in length.

Figure 4.3: Removing all video after 22 seconds.

4 Between the two **clips** add a cross-dissolve **transition** lasting three seconds.

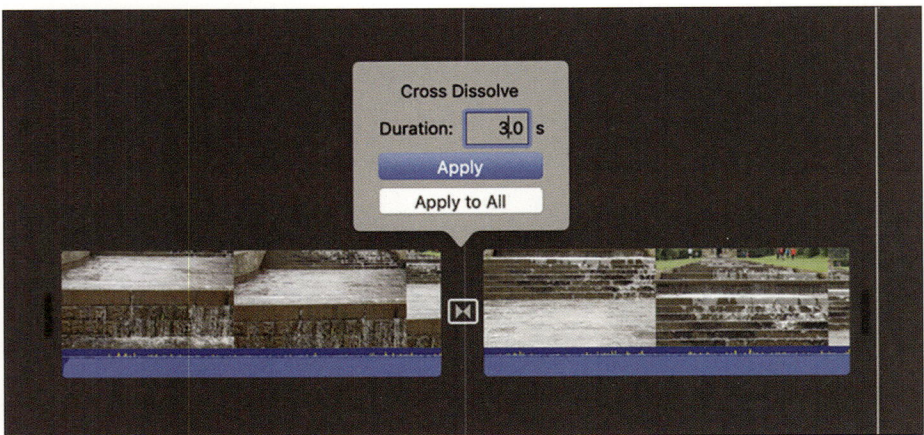

Figure 4.4: A cross dissolve transition being applied.

5 a Take a snapshot of the first **frame** of the video and insert it on the timeline at the start of the video so that it remains visible for four seconds.

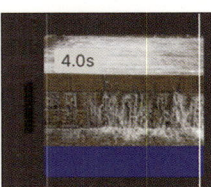

Figure 4.5: Still image taken from the video.

b Add the title 'the cascade' to this still image and a one-second wipe down transition between it and the start of the video.

Figure 4.6: Start of video with the title.

KEY WORDS

clips: short pieces of a video or audio file

transition: the method with which one video clip merges into a second clip

frame: a single image in a video file

6 The end of the video should fade to black over two seconds.

This can be done by adding another transition at the end of the video for a duration of two seconds.

Figure 4.7: Fade being added at end of video.

7 Use filters to make the first second of the video to be in greyscale. Then transition into colour with a one second cross blur filter.

This can be done by splitting the first video clip after one second, adding an effect to the clip to make it greyscale and then adding the transition.

Figure 4.8: Video in greyscale.

8 Export your video as an MP4 file with the title **Cascade_Finished**.

Practical tasks

Task 1

SKILLS
This task will cover the following skills: • importing new **tracks**   • trimming a sound clip to remove unwanted material   • normalising a sound clip.

KEY WORD

track: a specific recording, for example, of one instrument or voice. The tracks can then be edited separately and combined to play concurrently

Load the file **Singing1.mp3** into a suitable sound editing application.

This track was recorded in an outside environment with lots of people around.

In the introduction, there are instances of people talking and laughing.

1   Edit the sound track by deleting the first 22 seconds.

   The track is to be used in a video and is too long.

2   Edit the track so that it is exactly 1 minute and 3 seconds in length.

3   Normalise the track to a peak of −1.0 db.

4   Export your project to an MP3 file name **Chapter04_Task1_1**.

5   Load the file **Chapter04_Task1_1** and carry out the following effects:

   • Apply high pass and low pass filters to the entire clip. They should be set to a frequency of 1000 Hz and a roll-off (dB per octave) of 6dB.

   • Change the clip from stereo to mono.

   Export your project as a WAV file named **Chapter04_Task1_2**.

6   Load the file **Chapter04_Task1_1** and carry out the following effect:

   • Carry out equalisation on the entire file so that frequencies above 1000Hz are raised to 24 dB.

   Export your project as an MP3 file named **Chapter04_Task1_3**.

7   Apply a fade in and fade out to the start and end of **Chapter04_Task1_1**.

Export your project an MP3 file named **Chapter04_Task1_4**.

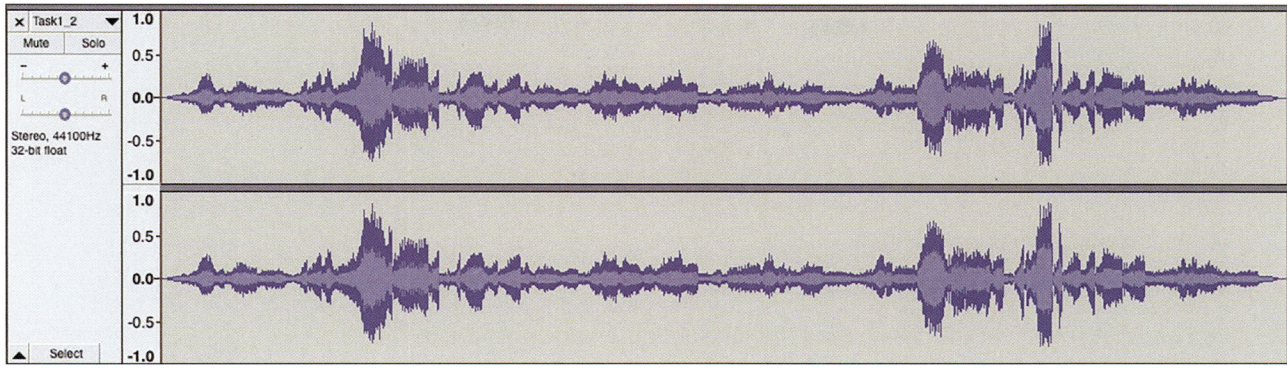

Figure 4.9: The track after editing.

8 **Load Chapter04_Task1_3** and apply noise reduction with the following parameters:

- Noise Reduction (dB) 12
- Sensitivity 6.00

Export your project as an AIFF file named **Chapter04_Task1_5**.

Task 2

The singer would like you to edit the clip to try to enhance the singing.

You have been asked to demonstrate the effect of different editing techniques on a part of the track.

1 Load your **Chapter04_Task1_2.mp3** file into a suitable sound editing application.

2 Copy a clip from the track from 7.521 seconds to 23.368 seconds.

3 Add a new track and paste this clip into it.

4   Paste in another copy leaving a short interval between them.

To this second clip apply reverberation with a reverberance of 100% and a room size of 75%.

5   Paste in a third copy leaving a short interval.

To this clip add an echo effect with a delay time of 100 and a decay factor of 10.

6   Paste in a fourth clip.

To this clip increase the pitch by 8%.

7   Add a fifth clip and increase the speed by 22.64%.

8   Add another track and in it paste the original clip at the same position as the fourth clip so that they will play together as a duet.

9   Finally, add another track and use this for a voice over to introduce each track.

These should be:

- as recorded
- clip with reverb
- clip with echo
- duet with clip with higher pitch
- clip with speed increase.

10   Export your clip as a WAV file named **Chapter04_Task2**.

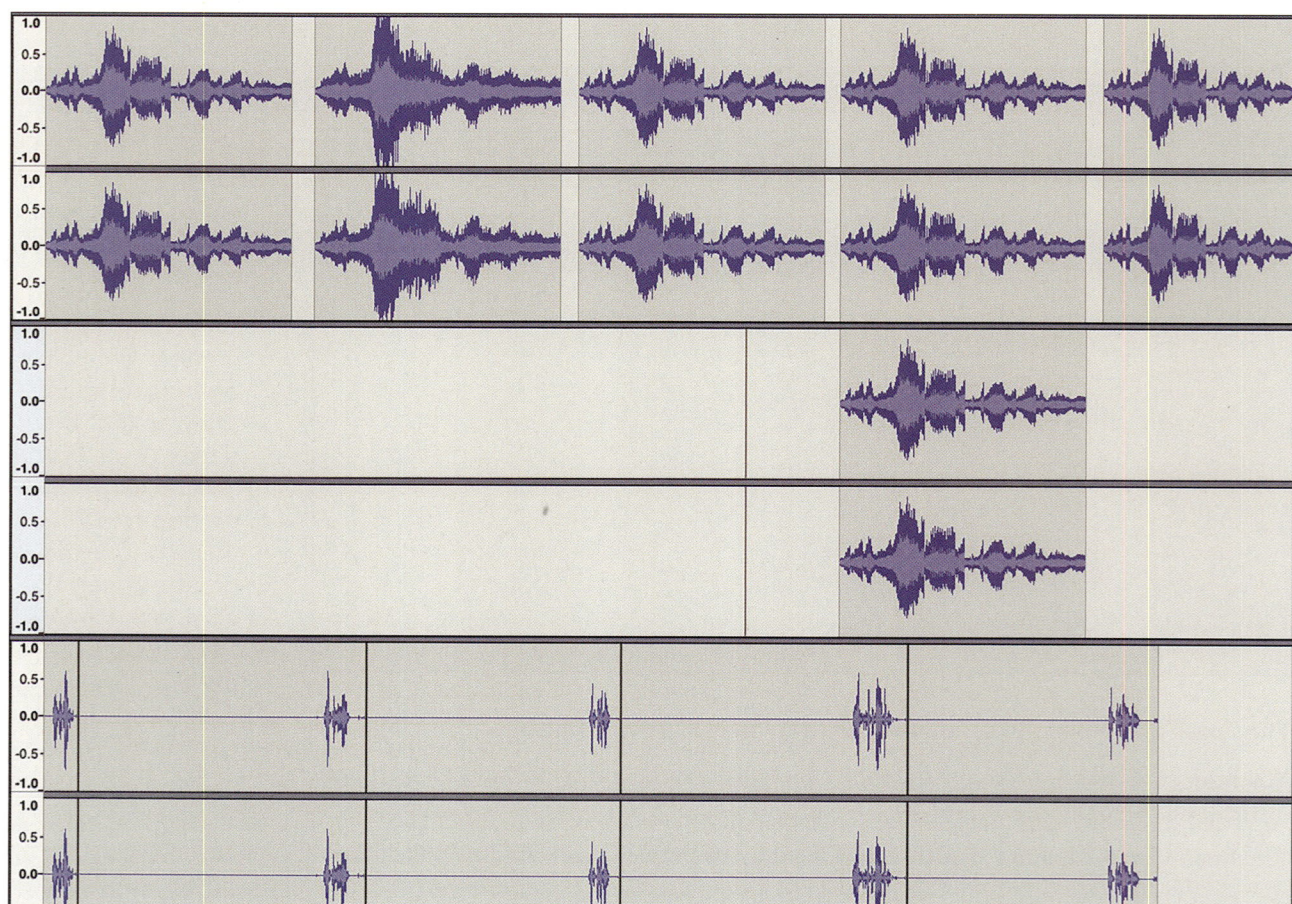

Figure 4.10: The tracks after editing.

Task 3

Load the file **Speech.mp3** into a suitable sound editing application.

This is part of a speech made by a politician in an election.

In it, he states, 'If you vote for me I will improve healthcare by employing more doctors, improve education by investing more money and I will raise more revenue by improving productivity.'

1 You have been asked to edit the sound clip in the following ways:

    a Remove all of the methods from the speech, i.e. remove 'by employing more doctors', 'by investing more money' and 'by improving productivity.'

    b Change the order of the speech to:

      'I will raise more revenue, improve education and I will improve healthcare if you vote for me.'

      You should cut the track where necessary and splice the sound clips together.

2 Export your finished work as an MP3 file named **Chapter04_Task3_Speech_2**.

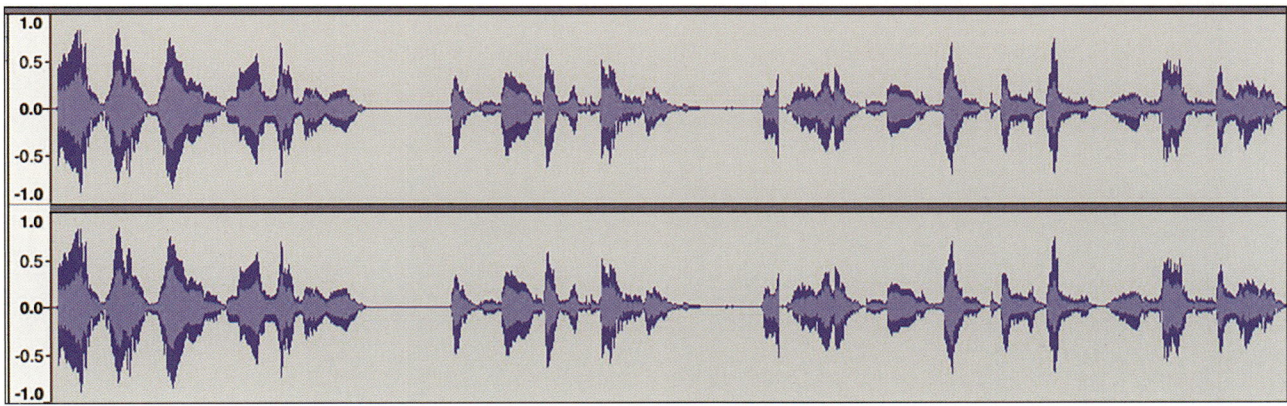

Figure 4.11: The speech after editing.

Task 4

Download the files **Paris1.jpg** to **Paris6.jpg**.

Your task is to create a video using these still images.

1   The specification for your video is:

- it should include the six images in their number order

- each video should be on screen for 10 seconds

- the images should pan and zoom

- there should be a cross dissolve between each image with a duration of 3 seconds

- your video should have an aspect ratio of 16:9.

2   Export your video project as an MP4 file with the name **Chapter04_Task4_ Paris_1**.

Figure 4.12: A cross dissolve transition between the images.

Task 5

1   You have been asked to develop your project in the following ways:

- Add the title 'Paris' to your first slide. This should be in a red font with a size of 234 point.

- For the other slides, there should be a scrolling subtitle stating their name:
  - Notre Dame
  - Louvre
  - Eiffel Tower
  - Champ de Mars
  - Place de la Concorde.

- After the last slide there should be a blank, black screen that reveals the text 'The end' for 3 seconds.

2   Export your video project as an MP4 file with the name **Chapter04_Task5_Paris_2**.

Figure 4.13: Title added to the first slide.

Figure 4.14: Scrolling subtitle on the last slide.

Task 6

SKILLS
This task will cover the following skill:
• adding sound to a video clip.

1   Edit your **Task1_1.mp3** file so that:
    • it is the exact length of the video
    • it fades in and it fades out.

2   Add this sound track to your movie.

3   Export your video project as an MP4 file with the name **Chapter04_Task6_Paris_3**.

4   Also export the video with an aspect ratio of 4:3 with the name **Chapter04_Task6_Paris_4**.

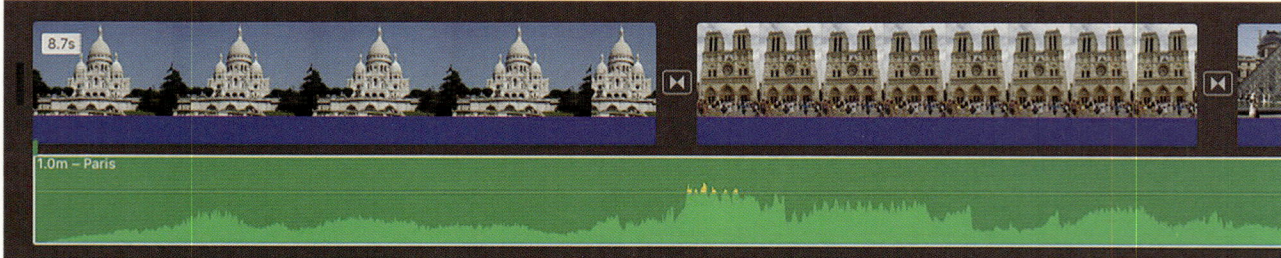

Figure 4.15: Video with attached soundtrack.

Task 7

1   Download the following video and audio clips:

- **Country_1.mp4**

- **Country_2.mp4**

- **Motorway.mp4**

- **Birdsong.mp3**

2   Add **Country_1.mp4** to the timeline and remove the audio track from it.

Figure 4.16: Country_1.mp4 added to the timeline.

Figure 4.17: Clip with audio track removed.

3   Add **Motorway.mp4** to the timeline with a one second wipe left transition between it and **Country_1**.

Increase the speed of this clip four times.

Figure 4.18: Wipe left transition and speed increase to **Motorway.mp4**.

4   Add **Motorway.mp4** to the timeline with no transition between it and the first copy.

Increase the speed of this clip four times and set it to play in reverse.

Figure 4.19: Clip added to play in reverse with speed increase.

5   Add **Country_2.mp4** to the timeline with a one second wipe right transition between it and the second **Motorway.mp4** clip.

6   Add the **Birdsong.mp3** audio clip to **Country_1** and adjust its length so that it is the same length as the video clip.

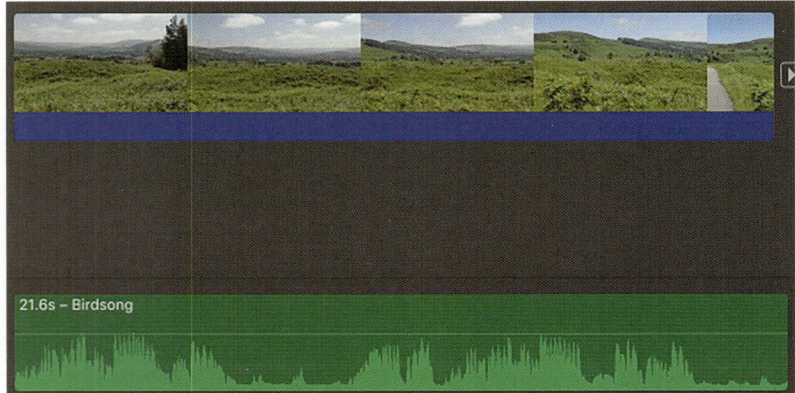

Figure 4.20: Birdsong.mp3 audio clip added to the timeline.

7   Add a black screen with the scrolling text 'Where would you rather be?' lasting eight seconds with a fade to black transition between it and the last video clip.

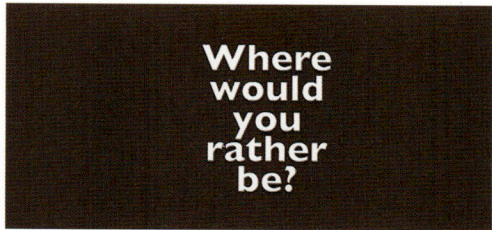

Figure 4.21: Black screen with scrolling text.

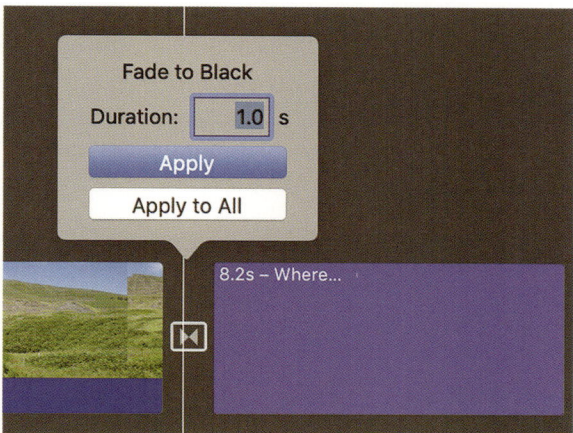

Figure 4.22: Fade to black transition added between clip and scrolling text.

8   Add the **Birdsong.mp3** audio clip to **Country_2** and the scrolling text, adjusting it to be equal in length to them.

Figure 4.23: Birdsong audio added to final clip and scrolling text.

9   Add scrolling credits, lasting for five seconds, at the end of the video. They should include the following credits: Camera, Sound, Editing, Publicity and Distribution. Also add in the names of the people responsible for each of these tasks.

10 Finally, export your movie as an MP4 file with the title **Chapter04_Task7_ Country_Motorway**.

Task 8

Your task is to create a video of the countryside.

1   Download and add **Country_3.mp4** to the timeline.

Cut the final 2.5 seconds from the end of this clip.

2   Add **Country_4.mp4** with a one second cross dissolve transition between them.

3   Take a snapshot of the frame at eight seconds of Country_4. Add it to start of the video so that it lasts for four seconds.

   a   Over the four seconds, the clip should change from a zoom view of the centre of the slide to a full view of it.

   b   Over the four seconds, the title 'The Peak District' should appear from blurred to clear at the bottom left of the screen.

   The title should be in a sans serif font with a size of 102 point.

4   After Country_4, add a black screen that should last for five seconds with 'The end' at the bottom left.

There should be a one-second fade to black transition between Country_4 and this black screen.

5   Download the file **Singing2.mp3** into a suitable sound editing program.

6   Carry out the following editing on this sound file:

   a   Remove the first 1 minute and 6 seconds of the clip.

   b   Now remove the end of the clip after 34 seconds.

   c   Normalise the clip with a peak amplitude of −1.0 dB.

   d   Add reverberance of 50% to the whole track.

   e   Export the file as **Singing2_Video.mp3**.

8   Add the file **Singing2_Video.mp3** to the timeline.

9   Export the video as both MP4 and MOV files with the titles **Chapter04_Task8_Scenery**.

10   As users may need to email this video, export the video with a resolution of 428 × 240 to a file named **Chapter04_Task8_Scenery_email**.

Export the file in the following formats: MP4, MOV and AVI.

Task 9

Your task is to insert a still image into the **Chapter04_Task8_Scenery.mp4** video that you created in Task 8.

1   Load **Still_Scenery.jpg** into your image editing software and edit the image to an aspect ratio of 16:9 to match the video. Export the new version as **Still_Scenery_2.jpg**.

Retain the portion shown below.

Figure 4.24   **Figure 4.24**

2 Insert **Still_Scenery_2.jpg** into the video **Scenery.mp4**.

It should be inserted into the timeline at 12 seconds and should stay on screen for 3 seconds.

Add a cross-fade transition between the clip before and the still image.

Figure 4.26

3 Export the video as an MP4 file named **Chapter04_Task9_Scenery_2**.

REFLECTION

- When you were learning how to use the sound and video editing software, did you prefer to start on your own, try out the features and learn by trying, failing and succeeding or do you prefer to use documentation, tutorials and guides?

- Which do you find the easiest to do, editing a video or a sound file? Can you explain why?

- When you are editing sound and video files, do you use the internet to research the techniques you should use?

SUMMARY CHECKLIST

- [ ] I can edit a video clip to meet the requirements of its intended application and audience.
- [ ] I know and understand why typical features found in video editing software are used.
- [ ] I can edit a sound clip to meet the requirements of its intended application and audience.
- [ ] I know and understand how and why typical features found in sound editing software are used.

Chapter 5

Mail merge

This chapter relates to Chapter 17 in the Coursebook.

The software used in this chapter is Word (Mac version).

LEARNING INTENTIONS

In this chapter you will learn how to:

- use/create/edit a source data file using appropriate software

- create a master document structure

- link a master document to a source file

- specify rules:

  - for selecting recipients

  - for managing document content

- set up fields:

  - for manual completion

  - for automatic completion

  - calculated fields

- use manual methods and software tools to ensure error-free accuracy

- perform mail merge.

Introduction

Mail merge increases efficiency and so saves time and money when similar information has to be communicated to many, often thousands or millions of people.

If most of the information in the communication is identical for each recipient, then it can be placed in a **master document** with spaces left where data specific for each individual can be placed. This individual-specific information or data can be kept in a data file such as a database or spreadsheet and the two are merged so that for each copy of the master document the specific data for one recipient is placed. This process occurs automatically with no user input unless it is specified in the master file.

KEY WORDS

mail merge: the automatic addition of data, such as names and addresses, from a source file into a master document, such as a letter

master document: the main document into which the data will be merged

WORKED EXAMPLE

The spreadsheet **Employees.xlsx** shows details about the tutors of North Eastern College.

The college has decided to reward employees who started working there before the year 2000 by giving them a pay rise of 10%.

Use the spreadsheet to create a mail merge letter to the employees who qualify. It should contain:

- the college letter head.

- a salutation with the employee's first and surnames.

- an explanation about the pay rise that contains:

  - when the person started working at the college

  - the number of years they have worked there

  - their current salary

  - their new salary after the 10% pay rise.

Save your master document as **MasterLetter.doc** and merge the letters to a file that you should save as **Letter.doc**.

The first task is to create a new document and link it to the data source, **Employees.xlsx**.

The letter can then be written, and the merge **fields**, First Name, Second Name, Salary and Start Year can be inserted using the Mailings menu.

Two **calculated fields** are required, one to calculate the number of years the employee has worked there and one to calculate the new salary.

To carry out the calculations, fields have to be added to contain the formulae required.

The formula for the number of years worked will be the present year minus the number in the Start Year merged field.

The formula for the new salary will be the current salary + 10%.

CONTINUED

These formulas and the merge fields are shown in Figure 5.1:

North Eastern College

A learning environment for everyone

{ DATE \@ "d MMMM yyyy" }

Dear { MERGEFIELD First_name } { MERGEFIELD Second_Name }

You started work at North Eastern College in { MERGEFIELD Start_Year } and have been here for { = 2019 - { MERGEFIELD Start_Year } } years.

To thank you for your service, we would like to offer you a pay rise of 10% and therefore your salary will rise from ${ MERGEFIELD Salary } to ${ ={ MERGEFIELD Salary } + ({ MERGEFIELD Salary }/10)}.

Yours faithfully

The Principal

Figure 5.1: Master document for the letter.

Before merging the letters, the data source must be **filtered** so that only those who started work before the year 2000 will receive one.

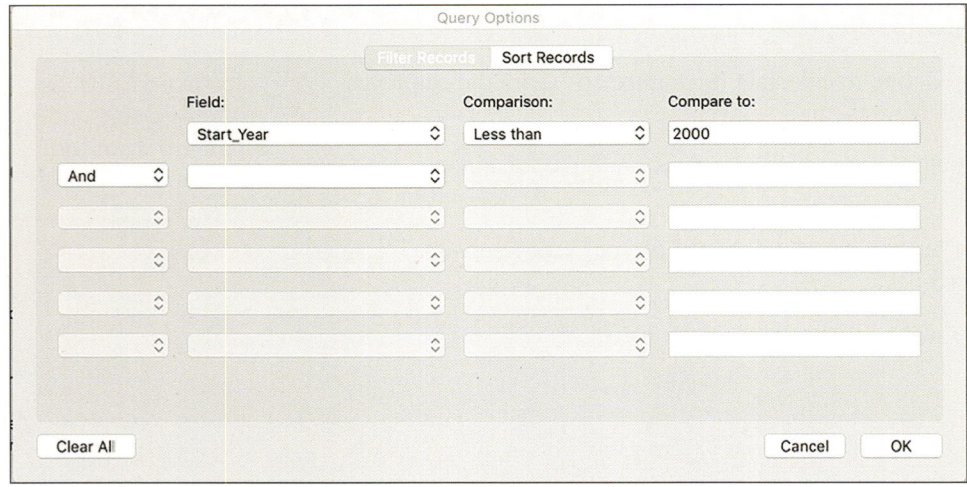

Figure 5.2: Required filter for the records.

CONTINUED

The letters can then be merged:

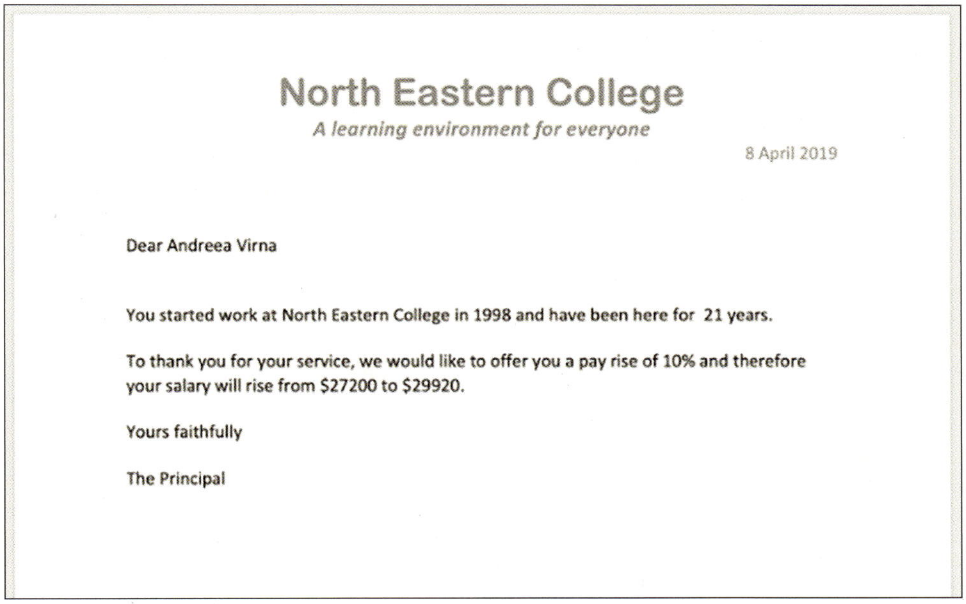

North Eastern College

A learning environment for everyone

8 April 2019

Dear Andreea Virna

You started work at North Eastern College in 1998 and have been here for 21 years.

To thank you for your service, we would like to offer you a pay rise of 10% and therefore your salary will rise from $27200 to $29920.

Yours faithfully

The Principal

Figure 5.3: The letter after merging with the data file.

These should be checked to ensure that only employees who started before 2000 receive a letter.

Practical tasks

Task 1

SKILLS
This task will cover the following skills:
• using source data
• creating a master document structure
• linking a master document to a source file
• identifying and using correct field names
• setting up a field for automatic completion.

KEY WORD

source file: the file containing the data that will be merged into the master document

North Eastern College educates students from the ages of 11 to 18. Each half term the college issues an assessment sheet for each student.

The data for some students is included in the data files:

- **Students1.csv**

- **Students1.xlsx**

1   Open the data file in a suitable application and look at the records showing some of the students at the college.

2   Create a mail merge letter with the following specifications:

- The name of the college with its mission statement 'A learning environment for everyone' and an automatic date field should be in the master document header.

- The letter should then show the full name of each student, their year and tutor group and a list of their subjects with their scores.

- The master document should be suitably formatted.

- Save your master document as **Chapter05_Task1**.

3   Use Finish & Merge/ Edit Individual Documents to check the merged documents.

A sample letter, using the data, is shown in Figure 5.4.

Figure 5.4: Master document for the letter.

North Eastern College
A learning environment for everyone

6 April 2019

Mid-term assessment

Student:	Catherine Byrne
Year and Tutor group:	7R1

Maths	63
English	61
IT	50
Science	40
Art	55

Figure 5.5: The letter after merging with the data file.

Task 2

SKILLS

This task will cover the following additional skill:

- specifying rules for selecting recipients.

The college would like to merge the assessment sheets for year 9 students only.

Use the following two methods to do this:

- filtering the recipient list
- using a Skip Record If rule.

Use Finish & Merge/ Edit Individual Documents to check that only year 9 reports are output.

Save the master documents as **Chapter05_Task2_Filter** and **Chapter05_Task2_Skipif**.

Task 3

SKILLS

This task will cover the following additional skills:

- specifying rules for selecting recipients
- specifying rules for managing document content
- setting up fields for manual completion.

As only the reports for year 9 students are to be printed out, the college would like the name of the head of year to be printed at the end of each report.

1   Using a Bookmark and ASK field, add a suitable head of year name.

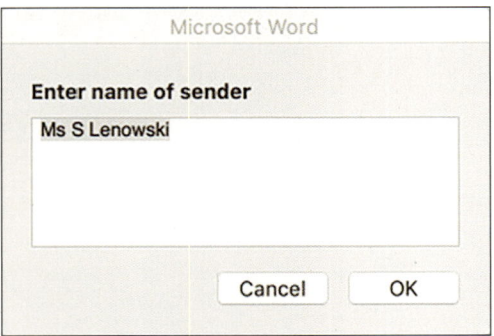

Figure 5.6: Inserted Ask field.

It can be set to just ask once, and head of year name is Ms S Lenowski.

Use Finish & Merge/ Edit Individual Documents to check that the head of year name is added to the reports.

Save the master document as **Chapter05_Task3_1**.

2   The college is really pleased with this addition and would now like you to add a FILL-IN field so that the head of year can add a comment for each student as they are merged.

To make it easier, the students should be sorted into alphabetical order.

Use Finish & Merge/ Edit Individual Documents to check that the correct comments have been added to the reports.

Save the master document as **Chapter05_Task3_2**.

An example is shown in Figure 5.7.

Figure 5.7: FILL-IN field for comments.

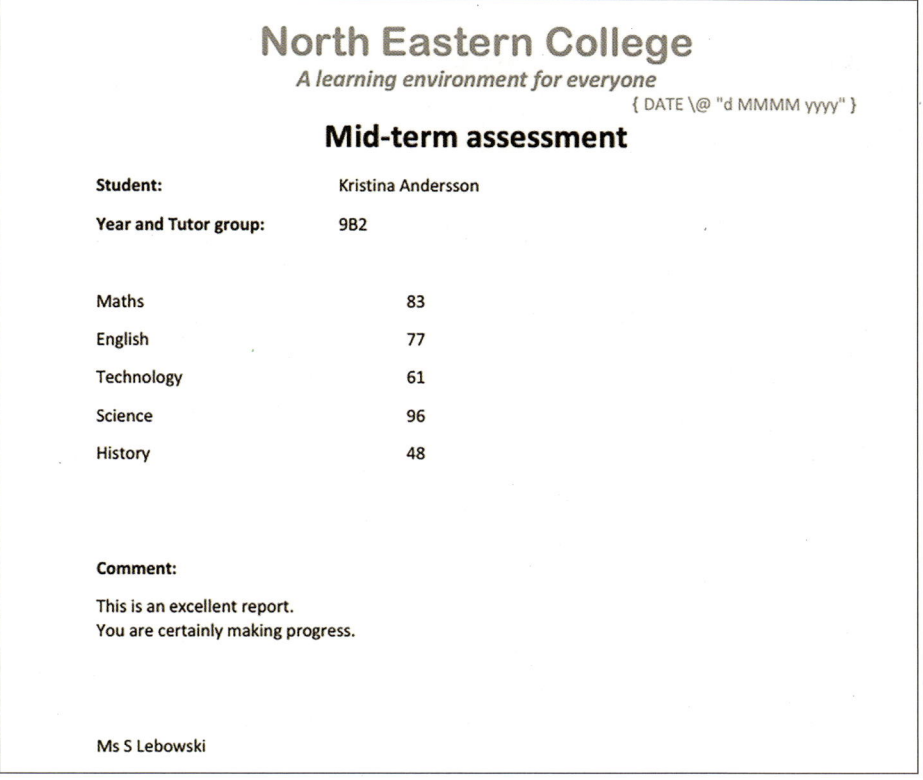

Figure 5.8: Report after merging and adding a comment.

The head of year is very pleased but has asked for one further improvement.

When they are asked to enter the comment, they would like the name of the student to be shown.

3   Display the field codes and add the merge fields (FirstName and Surname) to the FILL-IN field so that the head of year can see who they are writing a comment for.

Use Finish & Merge/ Edit Individual Documents to check that the correct comments have been added to the reports.

Save the master document as **Chapter05_Task3_3**.

An example is shown in Figure 5.9:

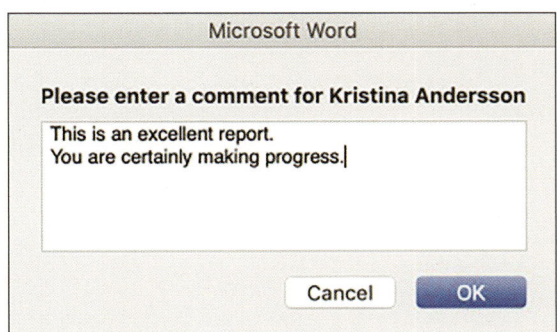

Figure 5.9: FILL-IN field with the name of the student.

Task 4

SKILLS

This task will cover the following additional skill:

- specifying rules for managing document content including conditional fields, IF THEN ELSE.

The heads of year would like to include messages congratulating students who do well in a particular subject.

Use conditional fields to insert messages saying 'Excellent' by each subject in which they obtain a score of 90 or more.

Use Finish & Merge/ Edit Individual Documents to check that the correct comments have been added to the reports.

Save the master document as **Chapter05_Task4**.

An example is shown in Figure 5.10:

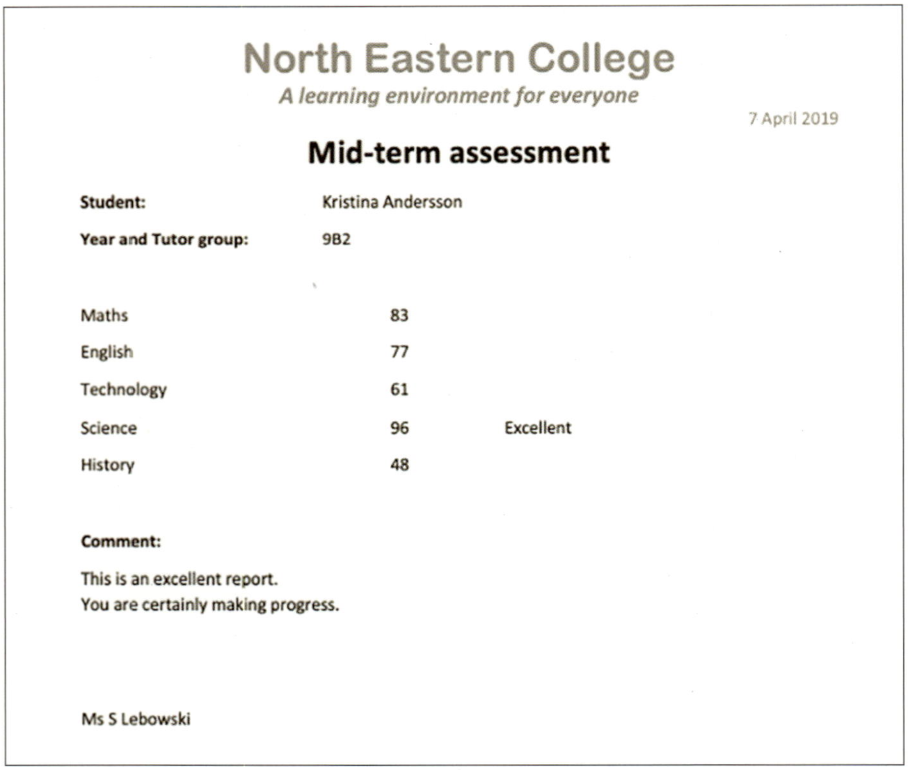

North Eastern College

A learning environment for everyone

7 April 2019

Mid-term assessment

Student:	Kristina Andersson	
Year and Tutor group:	9B2	

Maths	83	
English	77	
Technology	61	
Science	96	Excellent
History	48	

Comment:

This is an excellent report.
You are certainly making progress.

Ms S Lebowski

Figure 5.10: Report with a conditional field displaying a message if the score is 90 or more.

Task 5

Everyone is really pleased with your work on the reports.

And, of course, they would like just one more improvement!

They would like more comments by the subjects using the following rules:

90 or over	Excellent
70 to 89	Good
60 to 69	Satisfactory
Below 60	Must try harder

This can be done by adding another IF…THEN…ELSE field where the alternative text should go.

Just delete the two "" and insert another IF…THEN…ELSE field according to the rules shown above.

Use Finish & Merge/ Edit Individual Documents to check that the correct comments have been added to the reports.

Save the master document as **Chapter05_Task5**.

An example is shown in Figure 5.11:

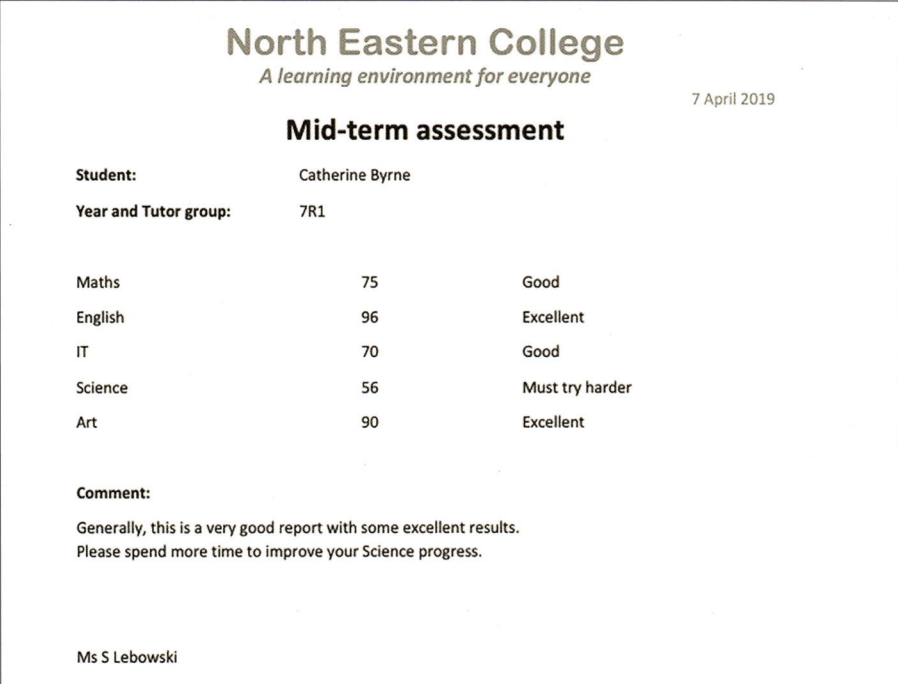

Figure 5.11: Report with conditional fields displaying different messages.

If your software does not support the use of 'nested loops' in mail merge, then you could add an extra column in the spreadsheet for the comment and then merge this data into your master document.

Task 6

SKILLS

This task will cover the following additional skill:

- creating calculated fields using arithmetical operators.

You are everyone's favourite person. The reports have never been so good.

But, of course, there is one more thing.

The heads of year would like the mean or average score shown on the report.

1 Place 'Average' as a side heading and then insert a field using a menu or key combination.

   The brackets should appear in the master document.

   { }

   Figure 5.12: The brackets used for entering a calculation field.

   The field should be toggled to show the field codes rather than the result.

   The formula can then be built up by inserting the required merge fields and arithmetic operators such as + and /.

   Don't forget to add an '=' symbol at the beginning and use brackets to do the additions before the division.

   Use Finish & Merge/ Edit Individual Documents to check that the averages have been added to the reports.

   Save the master document as **Chapter05_Task6_1**.

An example is shown in Figure 5.13:

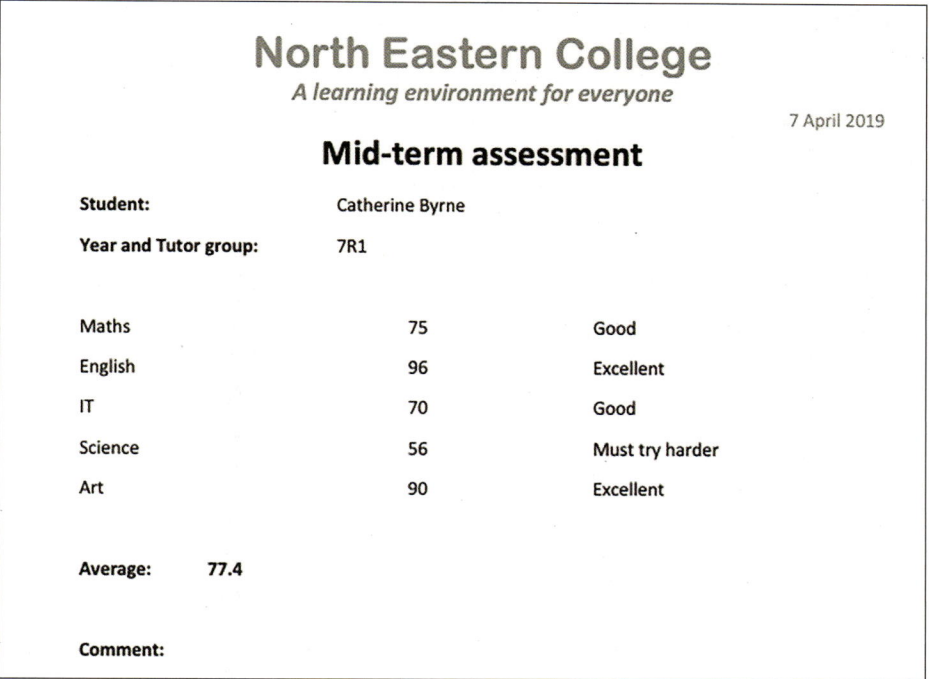

Figure 5.13: Report showing the result of the calculation of the average.

Just one more improvement. They would like the average without any decimal places.

2 Carry out research and add the correct commands to ensure that the average is an integer.

Use Finish & Merge/ Edit Individual Documents to check that the averages are integers.

Save the master document as **Chapter05_Task6_2**.

An example is shown in Figure 5.14:

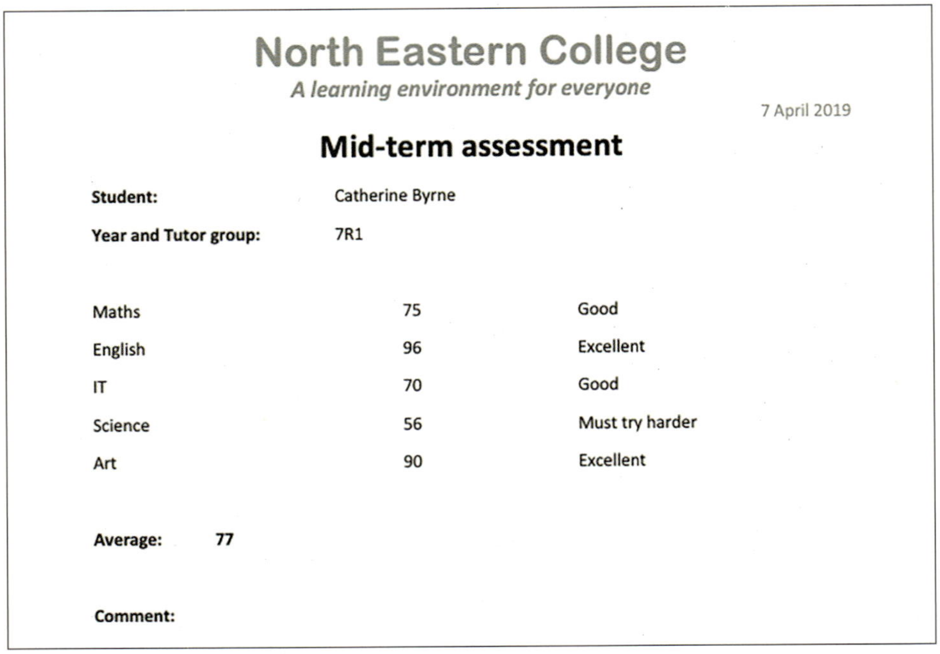

North Eastern College

A learning environment for everyone

7 April 2019

Mid-term assessment

Student:	Catherine Byrne
Year and Tutor group:	7R1

Maths	75	Good
English	96	Excellent
IT	70	Good
Science	56	Must try harder
Art	90	Excellent

Average: **77**

Comment:

Figure 5.14: Report showing the average as an integer.

Task 7

SKILLS

This task will cover the following additional skill:

- creating labels.

The reports are going to be distributed in sealed envelopes and the college would like you to create labels to stick on them.

The following fields should be on the labels:

- FirstName
- Surname
- Year
- TutorGroup

Create a mail merge document for the labels and add the fields requested.

Use Finish & Merge/ Edit Individual Documents to check that the labels are output as expected.

Save the master document as **Chapter05_Task7**.

An example is shown in Figure 5.15:

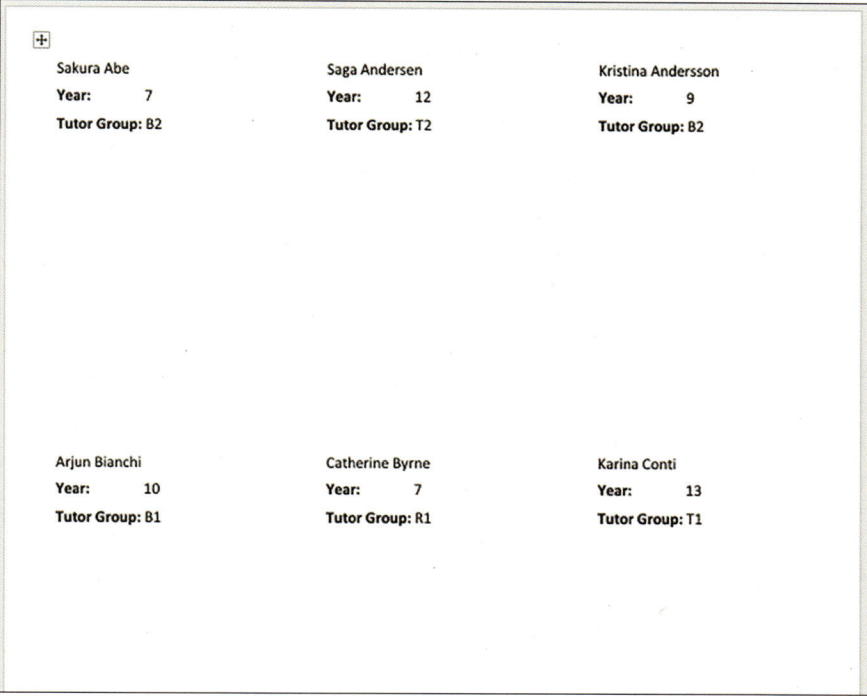

Sakura Abe
Year: 7
Tutor Group: B2

Saga Andersen
Year: 12
Tutor Group: T2

Kristina Andersson
Year: 9
Tutor Group: B2

Arjun Bianchi
Year: 10
Tutor Group: B1

Catherine Byrne
Year: 7
Tutor Group: R1

Karina Conti
Year: 13
Tutor Group: T1

Figure 5.15: Mailing labels.

Task 8

SKILLS

This task will cover the following additional skill:

- **embedding** a chart.

KEY WORD

embedding: importing data from a data source so that any changes to the data source are shown in the new document

The college has created a spreadsheet that, in addition to containing the student information, contains a chart showing the average mark obtained for each subject.

- Open the **Chapter05_Task1** document that you created earlier.

- Change its data file to this new spreadsheet – **Students_Chart.xlsx**.

- Copy the chart and embed it into the mail merge document so that there is a link between them.

- Check that it is actually linked by changing the data in the spreadsheet, e.g. by changing the IT average to 90 and checking that the change is reflected in the mail merge document.

Save your changed document as **Chapter05_Task8**.

An example is shown in Figure 5.16:

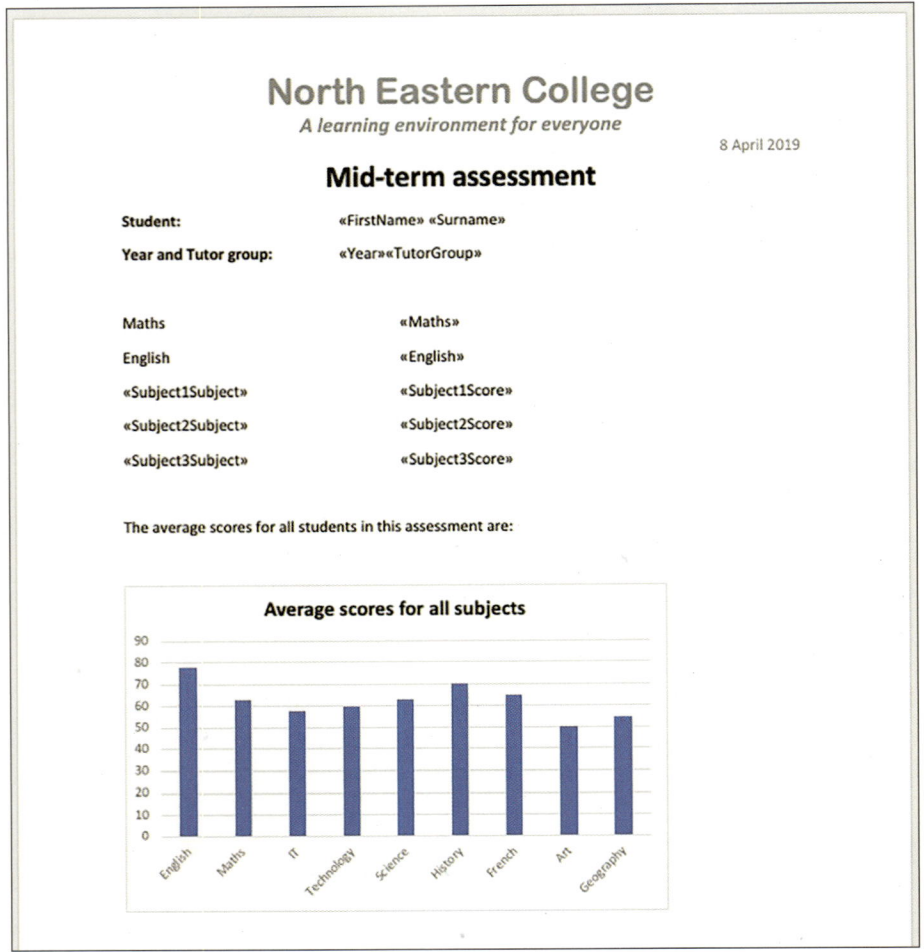

Figure 5.16: Report with an embedded chart.

Task 9

Open the spreadsheet **BigCorp.xlsx**.

It shows the sales figures for the employees of an international company called BigCorp.

The company would like you to produce the following:

- A mail merge letter to each employee informing them of their sales figures for each quarter of the previous year.

- The letter should also contain a chart showing the overall sales figure for each employee covering the whole year.

- The letter should also allow the managing director to add a comment on each individual letter.

- A mailing label for each employee showing their FirstName, Surname, Email and Area.

Save your documents as **Chapter05_Task9_Letter** and **Chapter05_Task9_Labels**.

Examples are shown in Figure 5.17 and Figure 5.18:

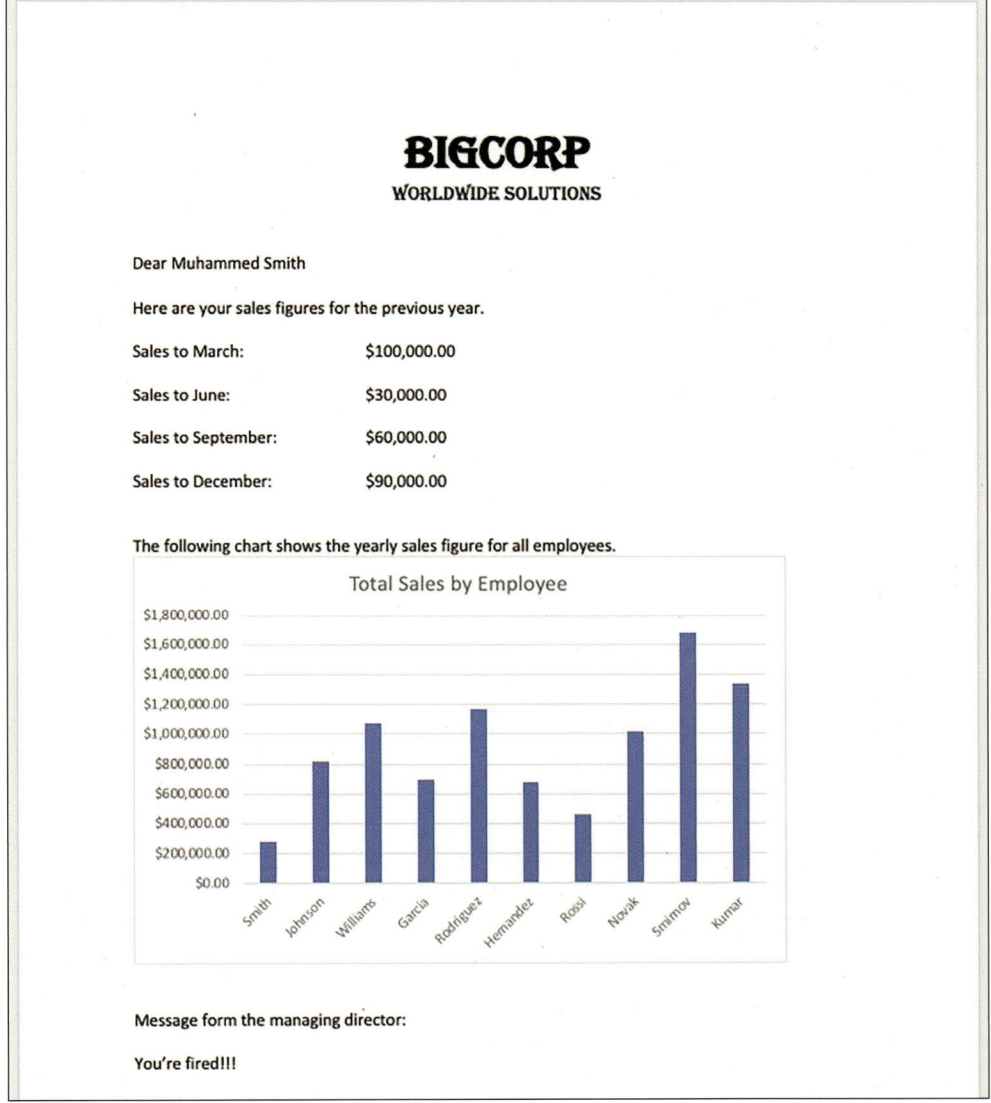

Figure 5.17: An employee letter with an embedded chart and comment.

Muhammed Smith	Emma Johnson	Jose Williams
msmith@bigcorp.com	ejohnson@bigcorp.com	jwilliams@bigcorp.com
North East	West Europe	South America
Maria Garcia	Nozomi Rodriguez	Lucia Hernandez
mgarcia@bigcorp.com	nrodriguez@bigcorp.com	lhernandez@bigcorp.com
Sweden	East Europe	Central America
Sophia Rossi	Tamar Novak	Alexei Smirnov
srossi@bigcorp.com	tnovak@bigcorp.com	asmirnov@bigcorp.com
Italy	South Africa	Russia

Figure 5.18: Mailing labels for the employee letters.

REFLECTION

- Can you think of any ways in which you could or would use mail merge in your everyday life?

- If your mail merge requires a calculation would it be easier to add a field in the master document or create a new field in the source document, e.g. a spreadsheet?

- Make a list of the stages you have to go through to create a mail merge document.

SUMMARY CHECKLIST

- ☐ I can use/create/edit a source data file using appropriate software.
- ☐ I can create a master document structure.
- ☐ I can link a master document to a source file.
- ☐ I can specify rules:
  - for selecting recipients
  - for managing document content.
- ☐ I can set up fields:
  - for manual completion
  - for automatic completion
  - calculated fields.
- ☐ I can use manual methods and software tools to ensure error-free accuracy.
- ☐ I can perform mail merge.

> Chapter 6

Graphics creation

This chapter relates to Chapter 18 in the Coursebook.

The software used in this chapter is Pixelmator Pro (Mac version).

LEARNING INTENTIONS

In this chapter you will learn how to:

- work with layers
- use transform, grouping and merging tools
- use alignment, distribution, layout, picker and crop tools
- change the opacity of an image
- create vector images to meet requirements of an audience
- use vector drawing tools:
    - selection
    - fill
    - node and path editing
- convert bitmap images into editable vector shapes
- create a bitmap image to meet the requirements of an audience
- use tools to:
    - select parts of an image
    - adjust colour levels
    - filter parts of an image
    - resize an image/canvas
- select font styles
- fit text to a path or shape
- set text in a shape
- convert text to editable vector shapes.

Introduction

Graphic images are an important medium of communication. We are surrounded by them in magazines, leaflets, advertising, logos and online. An image can create a mood or communicate excitement, sadness, reflection or an idea of perfection. They can be manipulated in many ways to enhance or change the impressions they create or meanings that they convey. As well as revealing truths, images can be manipulated to imply a situation completely different to what was originally intended.

WORKED EXAMPLE

A nursery, called The Rose Garden, as it specialises in roses, would like a circular logo that they can use on advertising material. They have provided you with an image of one of their flowers.

Figure 6.1: Image provided by the nursery.

A circular clip needs to be taken from the image, featuring the rose flower. This can be done using a clipping mask and the result should be saved as a .png file to preserve the transparent background.

Figure 6.2: Circular clip of the image.

A new **layer** can be created for a filled circle and this layer can be placed behind the image layer.

Figure 6.3: New layer with pink fill added behind the circular clip.

The name of the nursery can now be added in a suitable font. This can be done by fitting the text to a path or shape, or by converting the text to editable shapes which can be moved and rotated.

Figure 6.4: Image with path text added.

The circular fill can be softened using an effect called a Gaussian blur.

Figure 6.5: Fill after blur effect.

KEY WORD

layer: a 'surface' onto which an image or object is placed; each object is placed on a separate layer and they are stacked on top of each other (as though on different pieces of paper)

Practical tasks

Task 1

SKILLS

This task will cover the following skills:

- working with layers including moving back/front, merge and flatten
- using transform tools including skew
- using crop tools
- using group and ungroup
- using selection tools
- using distribution tools
- using fill tools
- using node and path editing.

Load **Task1.svg** into your graphics program. This consists of four shapes on separate layers.

Figure 6.6: Original image.

You have to edit the objects so that they appear as in Figure 6.7.

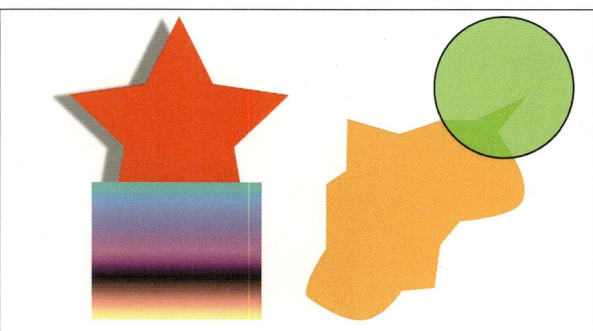

Figure 6.7: Image after editing.

To do this you will have to:

- Change the order of two of the layers.

- Add a shadow to the star.

- Change the fill of the square to a gradient with the colours shown.

- Transform the ellipse into a circle, add a 3-**pixel** stroke and change the fill **opacity** to 50%.

- Add points to the shape and edit it as shown.

    - Rotate and move the shape.

    - Change the fill colour to **RGB** 255 : 148 : 39.

Save your finished work as **Chapter06_Task1.jpg**.

Open file **Task1_2.svg**.

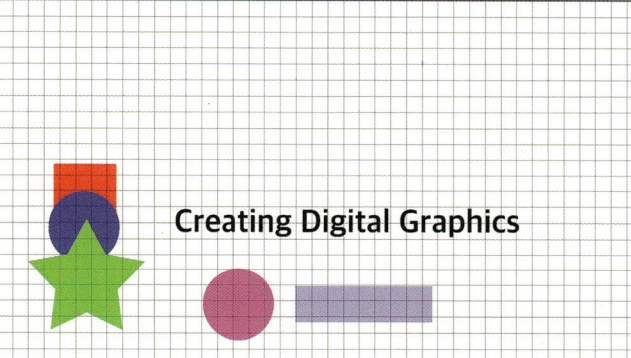

Figure 6.8: Original image.

Edit the objects so that they appear as in Figure 6.9.

Figure 6.9: Image after editing.

To do this, you will have to:

- Use the ruler and grid to align objects.

- Skew the text and align its left edge to a coordinate of 800 pixels.

- Group the three, overlapping object on the left, move them and change their alignments relative to each other.

KEY WORDS

pixel: a small square of one colour; these are combined to create a bitmap image

opacity: the lack of transparency of an image; at 0% opacity the image is fully transparent

RGB: red green blue colour system; all colours are a combination of quantities of red green and blue

- Group the ellipse and rectangle objects at the bottom, rotate them through 90° align the bottom edge to a Y coordinate of 1900 pixels.

- Unite the red square and blue ellipse and merge their layers.

- Rotate the star by 20 degrees.

Export your finished work as **Chapter06_Task1_2_Complete.jpg**.

Task 2

Your client has provided you with two images – **Task2_Garden.jpg** and **Task2_Cat.jpg**.

Figure 6.10: Image of garden.   **Figure 6.11:** Image of cat.

The client would like you to:

- Copy the cat, without its background, to the garden image, resize it and place it at the bottom left corner of the garden image.

- Make a copy of the cat, flip it horizontally and place the copy at the bottom right corner.

- Add the following text 'The Cat in the Zen Garden'.

- Convert the text to curves and edit their paths.

An example is shown in Figure 6.12.

Figure 6.12: Images after editing and adding text.

Save your finished work as **Chapter06_Task2** in the following formats: JPG, PNG, GIF, TIFF and PDF. Save each file separately.

Task 3

SKILLS
This task will cover the following additional skills: • using colour picker tools   • using transform tools   • using tools/filters to alter parts of an image   • changing the resolution of an image.

1   A client is disappointed with a picture of the Bridge of Sighs he has taken in Venice. Unfortunately, it is spoilt by a sign in the bottom left corner and some posts with red and white rope attached. Open file **Task3_Venice.jpeg** and remove the sign, posts and rope without affecting the background, as shown in Figure 6.14.

Figure 6.13: Image with sign, posts and rope.

Figure 6.14: Image after the sign, posts and rope have been removed.

Save your finished image as **Chapter06_Task3_1.jpg**.

2   The client would also like a copy of the new image with a motion swirl affect.

An example is shown in Figure 6.15.

Figure 6.15: Image with motion swirl effect.

Create the effects and save your image as **Chapter06_Task3_2.jpg**.

3   The client is also disappointed with all the dark and black colours and has requested a more artistic image where they are replaced by brighter colours.

Use the colour picker to select darker colours in the image and replace them for the client.

An example is shown in Figure 6.16.

Figure 6.16: Image with replacement colours.

Save your image as **Chapter06_Task3_3.jpg**.

4   As a final task, the client would like you to combine these edited images with a photograph of another view of Venice.

Open **Task3_Canal.jpeg** and add layers for the two images you have created.

Decide where to position them, how to rotate them and the opacity they should have.

An example is shown in Figure 6.17.

Figure 6.17: Images added as new layers.

Save your finished image as **Chapter06_Task3_4.jpg**.

5   The client would like to email this image to friends. Change the resolution so that the width is 640 pixels with the height changing in proportion.

Save your finished image as **Chapter06_Task3_5.jpg**.

Task 4

SKILLS
This task will cover the following additional skills: • using a clipping mask to select part of an image • fitting text to a path • converting a **bitmap** image into **vector** shapes.

KEY WORDS

bitmap: an image made up of small squares, called pixels; each individual pixel can only be one colour

vector: an image that uses geometric points and shapes; calculations are used to draw the image

1   A football club has sent you an image of their stadium and would like you to make a circular logo that they can use on letters and leaflets. They would like the text 'The Greatest Team in Football' to be around part of the circular image in the logo.

Load **Task4_Stadium.jpg** for the logo – or you could use an image of any other football stadium.

An example is shown in Figure 6.19.

Figure 6.18: Image of the stadium.   **Figure 6.19:** Circular logo made from the image.

Save your logo as **Chapter06_Task4_1.jpg**

2   So that the logo can be used at different sizes, the club would like you to convert it to a vector image.

Convert the image and save it as **Chapter06_Task4_2.svg**.

Task 5

Load **Task5_1.jpg**.

It is a holiday image. But it is of poor quality. The exposure, contrast and so on are all wrong and need editing.

Your task is to improve the image. First change the colour depth to 16 Bits/Channel and then amend the levels mentioned above.

An example of 'before' and 'after' is shown in Figure 6.20 and Figure 6.21.

Figure 6.20: Original image.

Figure 6.21: Image after editing.

Save your finished work as **Chapter06_Task5.jpg**.

Task 6

1. A client has sent you a photograph they took of an area of Berlin and would like you to remove the clutter of all of the cables so that it is a clearer and calmer image.

   Load the **Task6_Berlin.jpg** and using the repair and clone tools, edit the image to remove the cables.

   An example is shown in Figure 6.22 and Figure 6.23.

Figure 6.22: Original image.

Figure 6.23: Image with cables removed.

   Save your work as **Chapter06_Task6.jpg**.

2. The client also took an image of a friend but unfortunately the flash caused the friend's eyes to appear red.

   Load **Task6_RedEye.jpg**.

   Use your graphics software to correct this image and save it as **Chapter06_Task6_RedEye_Fixed.jpg**.

Task 7

A client has asked you to create a poster to highlight the need to protect our environment and planet.

They would like the following:

- Scenes of devastation or destruction shown in greyscale.
- Two children shown as black-filled shapes.
- Coloured balloons, the only coloured objects, to symbolise hope.
- The text 'Save our world'.

Two files have been provided for you to work with, although you can search for your own images.

Task7_Park.jpg shows two children who can be cut out and converted to shapes.

Task7_Background.jpg shows a scene of devastation.

You can use these images, or you could select some of your own.

An example is shown in Figure 6.24.

Figure 6.24: Final image after editing.

Save your work as **Chapter06_Task7.jpg**.

Task 8

A friend has taken some photographs at an outdoor exhibition of sculptures. The images of the actual sculptures are striking but the sky is very plain and boring.

Your friend has found another photograph with a dramatic sky.

Figure 6.25: The original image of the sculpture.

Figure 6.26: Image of the sky.

1 Open **Task8_Sculpture_1.jpg** and **Task8_Sky.jpg**. Replace the sky in the first image with that in the second. Pay special attention to the small area between the leaves of the trees.

2 Your friend would like a plaque adding to the plinth of the sculpture with its title – 'The Goddess'. The plaque should have circular cut-outs at the corners.

An example is shown in Figure 6.27.

Figure 6.27: Final image after editing.

Save your work as **Chapter06_Task8.jpg**.

Task 9

A heavy metal music fan has asked you to design a name plaque for their house, which is called 'Metallic'.

- It should have a steel / metal effect
- The depth of the plaque should be visible
- The letters should have a three-dimensional effect
- The four screws used to fix it to the wall should be shown.

An example is shown in Figure 6.28.

Figure 6.28: Example design for the name plaque.

> **TIP**
>
> Layers can be duplicated and slightly offset. Linear gradient fills can be used to show highlights on the letters.

Save your plaque as **Chapter06_Task9.jpg**.

REFLECTION

- When you were creating or editing images, did you save your work regularly or not until the images were complete?

- When you were thinking of ideas, did you search for similar images to give you inspiration?

- Did you use online tutorials for the graphics software you are using?

SUMMARY CHECKLIST

- [ ] I can work with layers.
- [ ] I can use transform, grouping and merging tools.
- [ ] I can use alignment, distribution, layout, picker and crop tools.
- [ ] I can change the opacity of an image.
- [ ] I can create vector images to meet requirements of an audience.
- [ ] I can use vector drawing tools.
- [ ] I can convert bitmap images into editable vector shapes.
- [ ] I can create a bitmap image to meet the requirements of an audience.
- [ ] I can use tools to:
    - select parts of an image
    - adjust colour levels
    - filter parts of an image
    - resize an image/canvas.
- [ ] I can select font styles.
- [ ] I can fit text to a path or shape.
- [ ] I can set text in a shape.
- [ ] I can convert text to editable vector shapes.

Chapter 7

Animation

This chapter relates to Chapter 19 in the Coursebook.

The software used in this chapter is Tumult Hype!
(only available on Mac).

LEARNING INTENTIONS

In this chapter you will learn how to:

- configure the stage/frame/canvas for animation
- import and create Vector objects
- control object properties
- use Inbetweening ('Tweening') tools
- set paths
- use layers
- control animations
- use animation variables when creating animations.

Introduction

'Animate' means to bring to life. When we create an **animation**, we are giving signs of life to **objects** we, or someone else, have created. We can make them move, change in shape and colour and grow larger or smaller. Animations are widely used online and in videos to present information for learning, entertainment, advertising, etc. and are a powerful medium of communication.

WORKED EXAMPLE

An organisation named Star Designs has asked for an introductory animation for their website featuring star shapes and moving text.

The specification is:

- The **stage** should be 600 px wide and 400 px high.
- The stage should have a grey background.
- Two stars with colourful, gradient fills should enter from the top over two seconds.

KEY WORDS

animation: a series of images that are played one after another to simulate movement

object: an image, or combination of images, that is manipulated as one item

stage: the area where the animation takes place, to be within the animation the object must be on the stage

CONTINUED

- The stars should then start to rotate, the left one clockwise and the right, anti-clockwise.

- Each complete rotation should take one second.

- The coloured letters of the word 'Star' should enter individually over two seconds until they are between the stars.

- The word 'Designs' also in coloured letters should then enlarge over two seconds from invisible until it is the same size as the other text.

- The stars should then stop spinning.

- The animation should be produced as an MP4 video and also as a web animation.

The length of the animation will be six seconds – two seconds for the stars to enter, two for the first text and two for the second text.

The only resources required are the two stars with gradient fills that can be produced in the animation software or imported from a graphics program. If they are bitmap images, then they should be saved as .PNG files to preserve their transparent backgrounds.

After creating the stage with the correct dimensions and background colour, the stars can be positioned for the start of the animation and in a **key frame** two seconds later.

The two stars with their motion paths over the two seconds are shown in Figure 7.1:

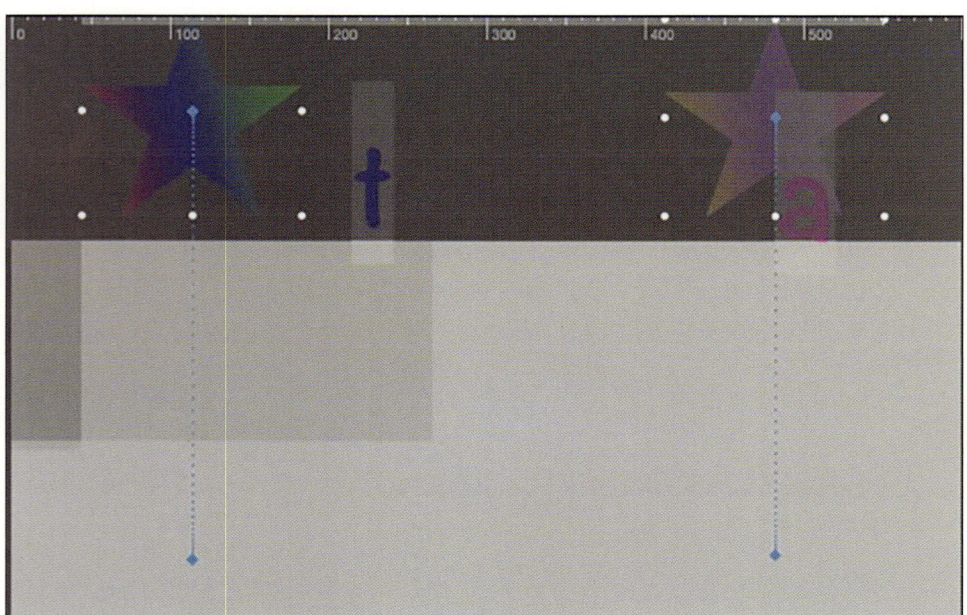

Figure 7.1: Two stars with motion paths.

The two stars will require a separate **timeline** for their rotations and this timeline will have to keep repeating over the next four seconds of the main animation timeline.

KEY WORDS

key frame: a location on a timeline that marks a frame that has a change in the animation, for example, a drawing has changed, or the start or end of a tween

timeline: the place that controls the order the frames are run, the positioning of the layers, and so on

CONTINUED

The letters of the word 'Star' must then appear over two seconds. Their motion paths are shown in Figure 7.2:

Figure 7.2: Letters with their motion paths.

The word 'Designs' must then appear over the final two seconds.

Figure 7.3: The completed animation.

The final design must then be exported as an MP4 video file and an HTML file so that it can be viewed directly in a web browser.

Load **WorkedExample.mp4** and **WorkedExample.html** to view them.

If you download **WorkedExample.html** you will also need the **WorkedExample. hyperesources folder**.

Practical tasks

Task 1

SKILLS
This task will allow you to practise some animation skills such as: • setting the size of the stage • creating vector objects • controlling object properties such as size, position, orientation and opacity • using **tweening** tools • setting paths and **layers** • controlling animations by looping animations.

KEY WORDS

tween: (Inbetweening) an animation where the start and end points are set; the computer generates the actual images to make the animation change

layer: an object or image given its own timeline for independent manipulation

coordinates: the position (x and y) of an object on the stage

1   Create an animation with the following elements and characteristics:

- The stage should be set at 600 px wide and 400 px in height.

- Create a star with a greyscale gradient fill. It should have the dimensions of 135 × 135 pixels and have the **coordinates** of left 128 pixels and top 102 pixels.

- Create a circle with a blue to red gradient fill. It should have a radius of 102 pixels and be placed at left and top 0 pixels.

Figure 7.4: Circle and star.

- Over three seconds the circle should move from the top left corner to the bottom right corner.

- Over the same three seconds, a rectangle should grow from the top left to the bottom right corner. As it grows it should change from a fill colour of #FF1CC8 to one of #F0DB82 and its opacity should change from 100% to 57%.

- Over the three seconds, the text 'Task 1' should grow from a font size of 0 to 64 px at the coordinates left 193 px and top 73 px.

- Over the three seconds, the star should rotate clockwise for one complete revolution.

Figure 7.5: Animation of circle, star and rectangle.

Figure 7.6: Final frame of the animation.

Save your animation as **Chapter07_Task1_1** in the native format of your animation software. Export it as **Chapter07_Task1_1.mp4** and in an HTML format if possible.

2   You should now make the animation repeat indefinitely as follows:

- At the end of the forward animation it should repeat in reverse so that all of the elements return to their starting positions, colours and opacities at the start.

But to make it slightly more difficult:

- The star should keep rotating in a clockwise direction even when the rest of the animation is running in reverse.

- The circle should morph into another shape.

Save your animation as **Chapter07_Task1_2** in the native format of your animation software. Export it as **Chapter07_Task1_2.mp4** and in an HTML format if possible.

Load **Chapter07_Task1_1.mp4**, **Chapter07_Task1_2.mp4**, **Chapter07_Task1_1. html** and **Chapter07_Task1_2.html** to view animation samples.

Task 2

Happy Smile play centre would like you to create an animated introduction for their website. They have provided the following specification:

- The canvas size should be 1024 px in width and 768 px in height.

- The background colour should be #F4C8EE.

- The letters for 'Happy Smile' should enter along individual motion paths over six seconds.

- The letters should be in different bright colours with a font size of 96 pixels.

- The word 'playcentre' should then appear beneath them from a font size of 0 to 64 pixels in one second. The font colour should be red.

- Over the seven seconds a red 'smile' should appear beneath them.

An example is shown in Figure 7.7:

Figure 7.7: Example of a completed animation.

Save your animation as **Chapter07_Task2** in the native format of your animation software. Export it as **Chapter07_Task2.mp4** and in an HTML format if possible.

Load **Task2.mp4** or **Task2.html** to view an animation sample.

Task 3

> ## SKILLS
>
> This task will cover the following skills:
>
> - using the ruler and grid to align objects
> - setting colour and size
> - using layers
> - controlling object properties such as position and opacity
> - using tweening to show motion
> - working with bitmap images.

A friend has some photographs that they took in Paris. They would like you to create a slide-show to display them. They would like:

- A stage size of 1024 px width and 768 px height. Make sure that the ruler and grid are visible.
- A background colour of #CBCDF0.
- The word 'Paris' to be shown first in a suitable font style and size. It should fade to an opacity of 0% over two seconds.

Figure 7.8: 'Paris' at 100% opacity.

- The first slide to then move in from the left over three seconds and remain displayed for two seconds.
- The second slide to then move in from the left as the opacity of the first slide changes to 0% over three seconds.
- All landscape and portrait images to be aligned to the same coordinates. Use the ruler and grid to do this.

Figure 7.9: Transition between two slides.

- This process should be repeated for all the slides, except for the last one that should appear from the right.
- Finally, 'Au revoir' in a red font should appear over three seconds on the last slide.

Figure 7.10: The final slide.

Download the images **Paris1.jpg** to **Paris6.jpg** from the **Task3 Images folder**, to use in your slideshow.

Save your animation as **Chapter07_Task3** in the native format of your animation software. Export it as **Chapter07_Task3.mp4** and in an HTML format if possible.

Load **Task3.mp4** or **Task3.html** to view an animation sample.

Task 4

SKILLS
This task will cover the following additional skill: •   controlling animations by looping or stopping them.

Your friend is very pleased with the slideshow animation, but they have requested one further feature.

•   They would like two buttons added so that they can pause it, to study the images in more detail, and then continue the animation.

Your task is to add and test these buttons.

Figure 7.11: Added pause and restart button.

Save your animation as **Chapter07_Task4** in the native format of your animation software. Export it as **Chapter07_Task4** in an HTML format.

Load **Task4.html** to view an animation sample.

Task 5

Your local garage has provided you with an image of the latest model of a car range that they sell.

They would like you to create a display animation with the following features:

- The stage should be 1240 px width by 800 px height.

- A black line with a thickness of 19 px, to act as a road, should appear from the left over two seconds. It should have a Y coordinate of 500 pixels.

- The car, cut out from the image provided, should then drive along the line from offscreen-left to offscreen-right over five seconds.

- To make it more realistic, they would like the wheels to rotate independently.

- As the car is moving from left to right, the wheels should rotate in a clockwise direction with one complete revolution every second.

- The car should then reverse from right to left over five seconds with the wheels now rotating in an anti-clockwise direction.

- Over these ten seconds, two smaller versions of the car should move, one from left to right and one from right to left.

- The cars should follow undulating motion paths and rotate to follow the paths.

- Their wheels should be rotating in the correct directions.

- The animation should repeat indefinitely.

Download **Task5.jpg**.

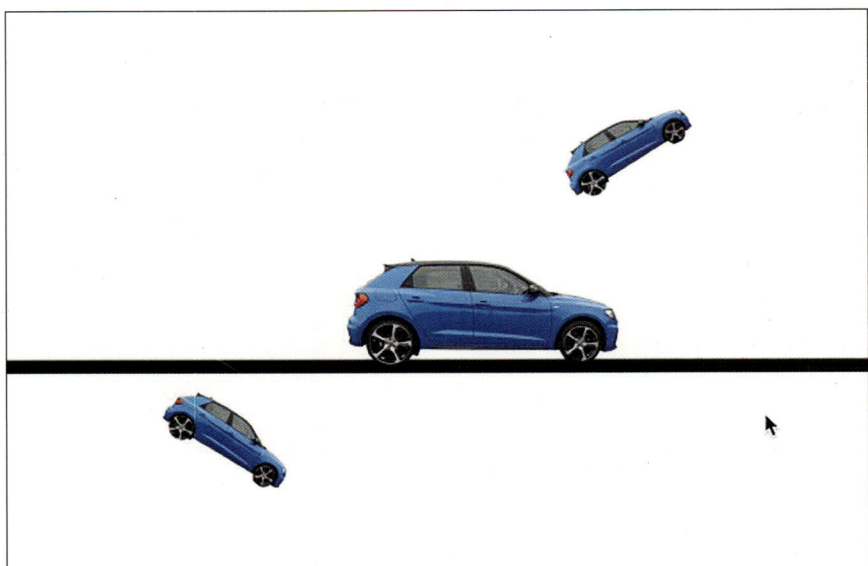

Figure 7.12: Example animation.

Save your animation as **Chapter07_Task5** in the native format of your animation software. Export it as **Chapter07_Task5.mp4** and in an HTML format if possible.

Load **Task5.mp4** or **Task5.html** to view an animation sample.

Task 6

SKILLS

This task will allow you to practise your animation skills including:

- controlling objects – size and position

- setting paths.

The Computer Museum would like you to create an animation to illustrate the exciting times of the 1980s when there were many different types and makes of computer.

They have provided you with eight named images of the computers that can be downloaded from the **Task6 Images folder**. They have also provided the following specification:

- The stage should be 800 px in width and 600 px in height.

- The heading should be 'Computers of the 1980s'. Use the ruler and grid to ensure it is centred horizontally.

- The following text should be used for the introduction:

A long time ago

in a galaxy far, far away...

Actually it was

on earth

In the 1980s.

There was an explosion

in the manufacture of computers.

Figure 7.13: Introduction text.

TIP

The word 'explosion' here means 'huge increase'.

- To make it more interesting, they would like the text to scroll up the screen, over seven seconds, like the introduction to the Star Wars film.

A long time ago
in a galaxy far, far away...

Actually it was

on earth

Figure 7.14: Scrolling animation.

A long time ago
in a galaxy far, far away...

Actually it was

on earth

In the 1980s.

There was an explosion

in the manufacture of computers.

Figure 7.15: Scrolling animation.

- Each computer image should then:
  - grow from a zero size as it moves along a motion path to the centre of the screen over one second
  - stay in this position for one second
  - move along a different motion path to its original starting position and diminish to zero size over one second.

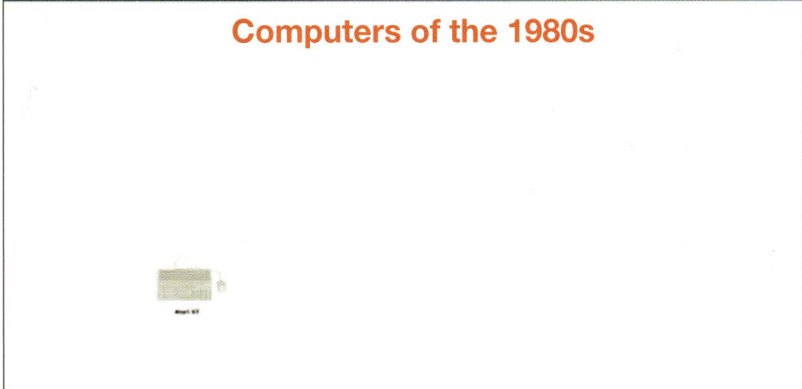

Figure 7.16: Start of computer motion path.

Computers of the 1980s

Atari ST

Figure 7.17: End of computer motion path.

- When all the images have appeared and disappeared, the text 'THE END' should move in from the left to the centre of the screen over two seconds.

  Save your animation as **Chapter07_Task6** in the native format of your animation software. Export it as **Chapter07_Task6.mp4** and in an HTML format if possible.

  Load **Task6.mp4** or **Task6.html** to view an animation sample.

Task 7

In Chapter 4 of the Coursebook, you learnt about algorithms and flowcharts.

This task will allow you to use your animation skills to produce an educational animation to help students learn about flowcharts and iteration.

- The animation should compare two numbers to see if they are equal.

- If they are not, then the animation should show the algorithm looping to select another number to compare.

- If they are identical, the flowchart should move to the end.

- The animation should cycle twice – once with different numbers and once where they are identical.

The images in Figures 7.18 to 7.21 are from a sample animation:

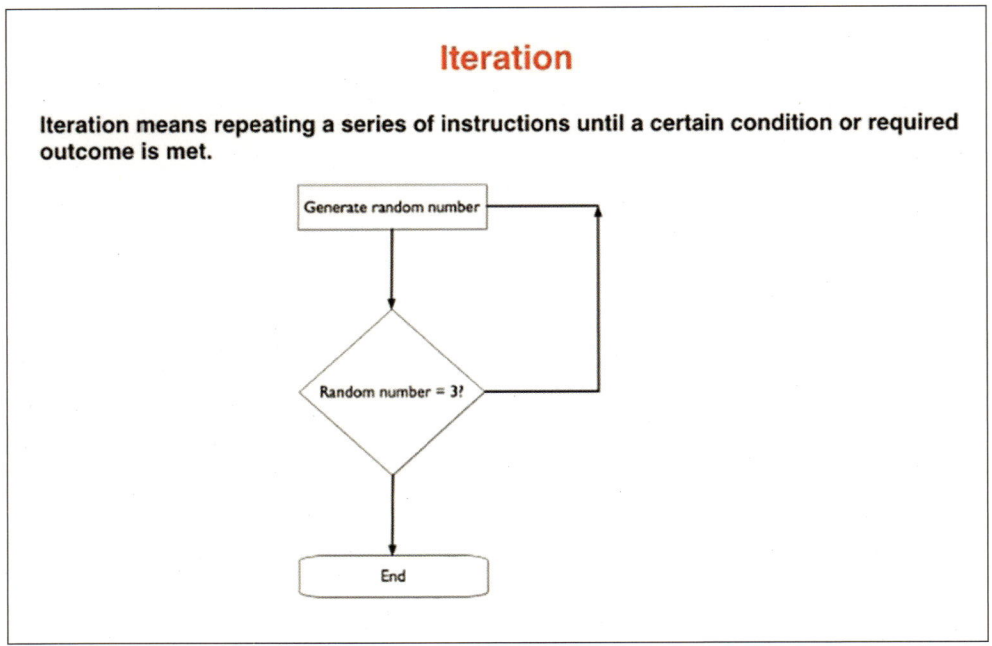

Figure 7.18: Start of the animation.

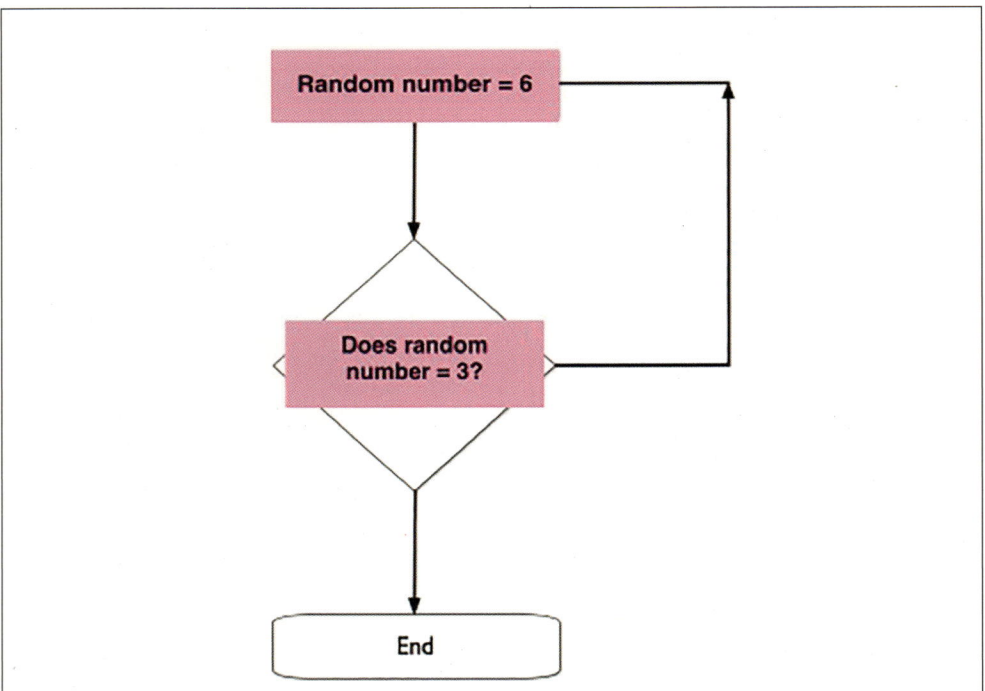

Figure 7.19: Animation comparing numbers.

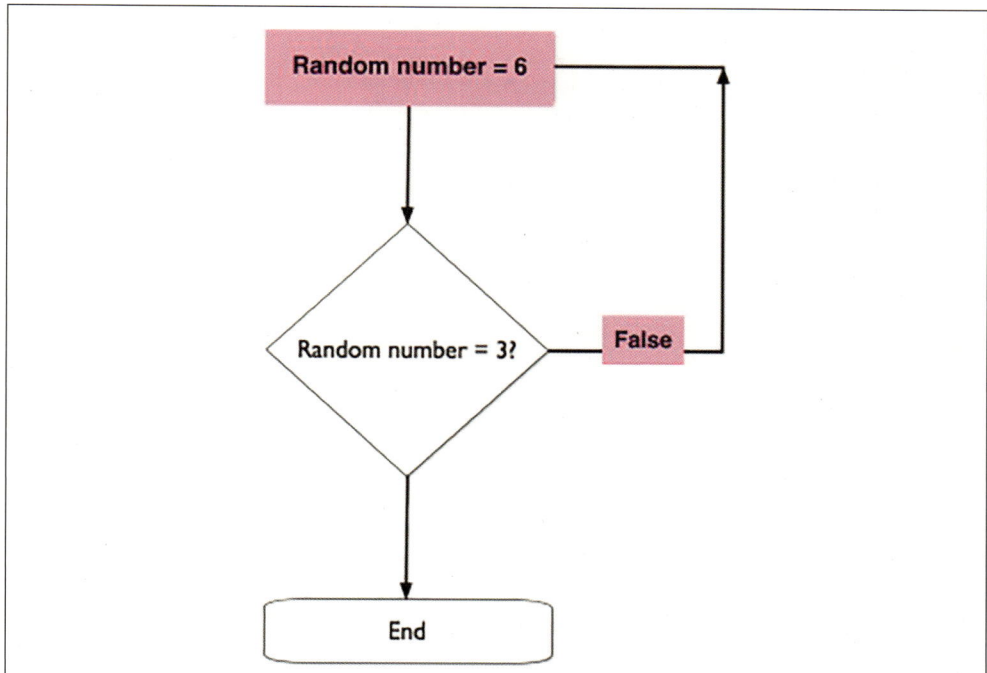

Figure 7.20: Animation showing the numbers are not equal.

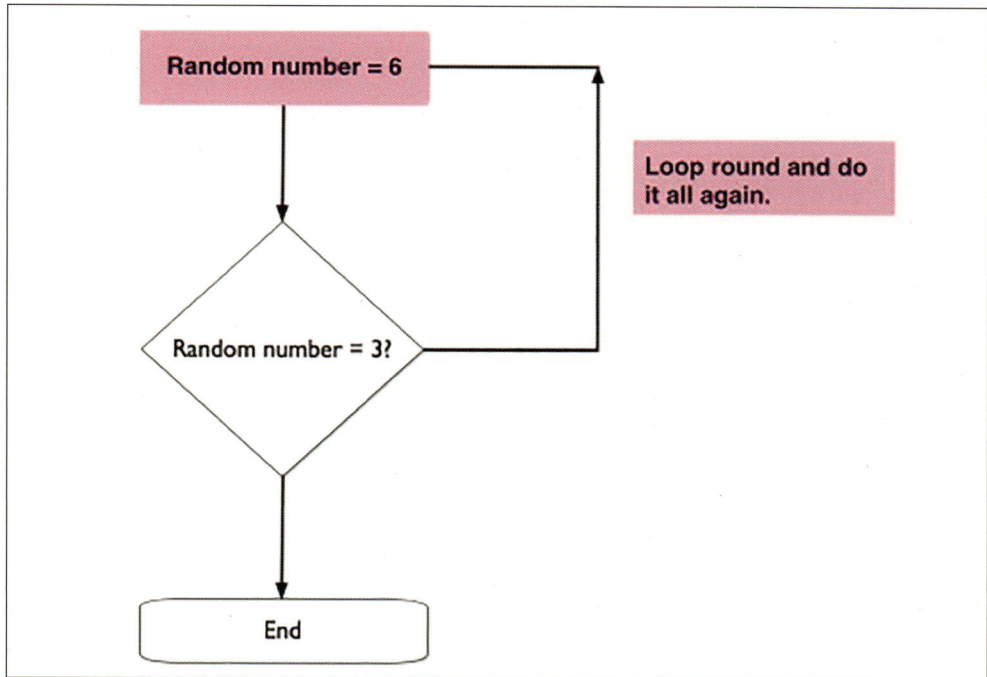

Figure 7.21: Animation looping back to generate another random number.

Save your animation as **Chapter07_Task7** in the native format of your animation software. Export it as **Chapter07_Task7.mp4** and in an HTML format if possible.

Load **Task7.mp4** or **Task7.html** to view an animation sample.

Task 8

SKILLS

This task will allow you to use your animation skills to produce a more sophisticated educational animation to help students.

In task 7, you were creating an animation about something you had studied during the course, but often educational animators have to produce animations about concepts they have not studied before.

In this animation you will be creating an animation about a method computers use to sort numbers into ascending order. It is called bubble sort and is not very sophisticated, but computers can get away with it because they are so fast.

For example, if there are the following numbers

7 3 9 5 2

The computer starts at the left and compares the first two numbers – 7 and 3 – and switches them so that they are ascending order:

3 7 9 5 2

It then compares the next pair – 7 and 9 – and switches them if necessary.

Therefore, at the end of the first pass the numbers will be:

3 7 5 2 9

It then starts Pass 2 and laboriously starts from the left comparing pairs of numbers so that at the end it will be:

3 5 2 7 9

Then it will start Pass 3 and continue to Pass 4 – whoever said that computers were clever?

Your task is to create an educational animation to show this.

- The numbers 4 5 3 2 7 1 6 should be sorted into ascending order.
- The animation should show the name of each pass and show the numbers being swapped, if necessary.
- There should be explanatory text to explain what is happening.

An example is shown in Figure 7.22:

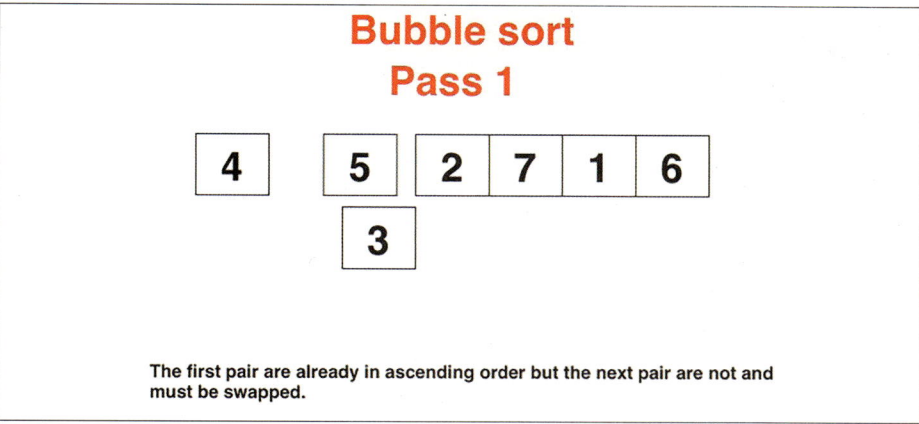

Figure 7.22: Animation showing the exchange of numbers 3 and 5.

Save your animation as **Chapter07_Task8** in the native format of your animation software. Export it as **Chapter07_Task8.mp4** and in an HTML format if possible.

Load **Task8.mp4** or **Task8.html** to view an animation sample.

Task 9

SKILLS

This task will allow you to create vector objects in addition to using your other animation skills. As you create and align the separate objects, you should use the ruler and grid.

This task will allow you to create vector objects in addition to using your other animation skills. As you create and align the separate objects you should use the ruler and grid.

The following face was made out of ellipses and rounded rectangles.

Figure 7.23: Vector graphic face made of separate objects.

Figure 7.24: The eyes, ears, eyebrows and mouth during animation.

An advantage of creating an overall image out of separate vector graphics is that they can all be animated separately.

Your task is to create a face and an animation so that the eyes, eyebrows, ears and mouth, at least, move independently. You can do more, if you wish.

Save your animation as **Chapter07_Task9** in the native format of your animation software. Export it as **Chapter07_Task9.mp4** and in an HTML format if possible.

Load **Task9.mp4** or **Task9.html** to view an animation sample.

REFLECTION

- Before you create an animation, do you plan how it will start, progress and end or do you make a start and then just let it develop? If you have used both methods, which produces the best results?

- When you are learning how to use the software, do you prefer to start on your own, try out the features and learn by trying and failing and succeeding? Or do you prefer to use documentation, tutorials and guides?

- If you use help, do you prefer written guides and tutorials or educational videos and animations?

SUMMARY CHECKLIST

- [ ] I can configure the stage/frame/canvas for animation.
- [ ] I can import and create Vector objects.
- [ ] I can control object properties.
- [ ] I can use Inbetweening ('Tweening') tools.
- [ ] I can set paths.
- [ ] I can use layers.
- [ ] I can control animations.
- [ ] I can use animation variables when creating animations.

Chapter 8

Programming for the web

This chapter relates to Chapter 20 in the Coursebook.

The software used in this chapter is an integrated envelopment environment (IDE), Brackets.
Any IDE or text editor can be used for the tasks.

LEARNING INTENTIONS

In this chapter you will learn how to:

- add interactivity to web pages
- change HTML content and styles
- show/hide HTML elements
- display data in different ways
- react to common HTML events
- provide user interaction
- create statements
- use JavaScript loops for iterative methods
- create functions
- use JavaScript timing events
- add comments to annotate and explain code
- using iteration.

Introduction

In this chapter, you will use and develop your programming skills to add interactivity to web pages to bring them to life. Web pages are coded using HTML and CSS. These are scripting languages used to design web pages and instruct browsers how and where elements should be displayed on a page. They are not programming languages and so JavaScript is currently used to allow users to interact with web pages. When you select an item, enter your password or order a new computer, the web page is using JavaScript programs to allow you to perform those actions. JavaScript uses the same basic constructs as other languages including sequence, **selection** and **iteration**. Like all programming languages it allows you to think logically and use your creativity to solve problems. For all problems there can be many solutions, but some are more elegant and efficient than others. Some suggested ones are given for the tasks.

WORKED EXAMPLE

You have been asked to create an element for a web page that allows a user to enter a password that is then checked. The user should be notified if it is recognised or not and after three unsuccessful attempts it is locked.

To solve this problem there must be the following:

*   an input box on the page for password entry

*   a store of recognised passwords

*   a **function** to compare the password entered with stored passwords with an action to call the function

*   messages for the user

*   a method of preventing more than three attempts.

An input box can be used for password entry and a button to call the function. In this instance, an onclick event of a button is better than an onchange or oninput event of the input box.

```
<body>
 <p style="font-family:arial; font-size:110%;">Please enter your password.</p>

    <input type = "text" id = "entry" style="font-family:arial; font-size:110%;"> <button style="font-
    family:arial; font-size:110%; color:red" onclick = "response()">Submit</button>

</body>
```

Figure 8.1: HTML code for paragraph and input box.

The input box is given an ID and the button calls a function named response().

The JavaScript code is contained within `<script></script>` tags either in the head or the body of the page.

Three **global variables** are defined so they can be accessed by different functions.

```
<script>
   // The following global variables are declared
   var found = "no";
   var attempt = 0;
   var passwords = ["password1", "password2", "password3"] //an array to store recognised passwords.
```

Figure 8.2: Declaration of variables.

Conditional operators and a loop are required to compare the entered password with those stored in the **array** and decide it is the same as one of them.

All code should have comments to explain the logic of the program.

KEY WORDS

selection: use of a conditional statement to decide a course of action or which section of code to run

iteration: a loop to repeat a section of code for a fixed number of times or until a required outcome is achieved

function: a separate piece of code that has an identifier and performs a task, it can be called from elsewhere in the code and can return a value

KEY WORDS

global variable: a variable declared outside a function; it can be accessed anywhere within the code

array: a data structure that can store multiple items under one identifier; the items are of the same type

```
function response() {
    attempt = attempt + 1;  // each time attempt is incremented by 1.

    if (attempt < 4 && found == "no"){  //the following loop will be carried out only if attempt is less
    than 4 and the correct password has not been added.


    var thisPassword = document.getElementById("entry").value;// the password is obtained from the input
    box.

        for (x = 0; x < passwords.length; x++){ //this loop will check the password entered with all the
        passwords in the array.

            if(thisPassword == passwords[x]){ //if the password is recognised this variable is changed to
            'yes' so that thius code will not be run again.
                found = "yes";

            }
        }
            if (found == "yes"){    //user message if password is recognised.

                alert("Password recognised.");
                return; // there could be a command here to break out of this function to another one if
                the password is correct.
            }

            else {
                alert("Sorry password not recognised. Please try again.");

            }

    }
    if ( attempt == 3) {
        alert("Sorry but the password is now locked."); // message to inform user that it is now locked.
    }
}
```

Figure 8.3: JavaScript code for checking entry and presenting a response to the user.

Practical tasks

Task 1

SKILLS

This task will cover the following skills:

- changing HTML content

- changing HTML styles – colour

- reacting to common HTML events – `onclick`

- creating statements – variables, conditional

- creating functions – executed when an event occurs

- using conditional operators – `if, else`

- displaying data in different ways – writing into an alert box

- interaction – `prompt()`.

1   You have been asked to create user interaction on a web page to allow a user to change the font colour in a particular paragraph.

The user enters the colour and the font colour is changed when a button is pressed.

An example is shown in Figure 8.4 and Figure 8.5:

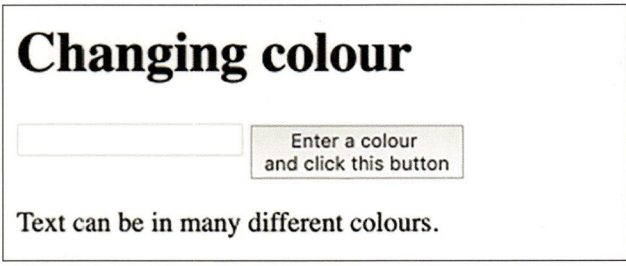

Figure 8.4: Entry box and button.   **Figure 8.5:** Text after colour selection.

To complete the task you will need:

- a heading for the page

- a paragraph for the text (don't forget to give it an ID)

- a text box where a user can input the required colour

- a button with an onclick event to change the text colour of the paragraph

- a function, called by the button, which will obtain the colour from the text box and change the colour property of the text in the paragraph.

Save your file as **Chapter08_Task1_1.html**.

2   You should now add an element to notify the user before the colour is changed.

Figure 8.6: User notification.

To complete the task you will need to:

- add a command for an alert message to be displayed before the colour is changed.

Save your file as **Chapter08_Task1_2.html**.

3 A further improvement has been requested. Instead of the alert you have to use a confirm box so that the user can click on 'OK' to make the colour change or 'Cancel' to abort the change.

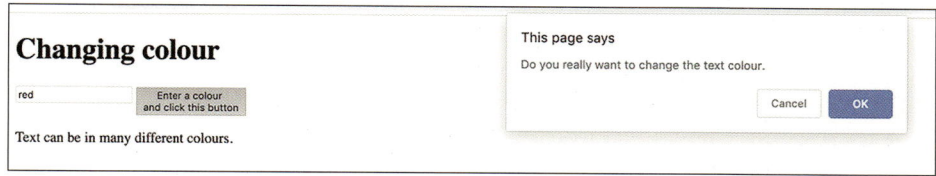

Figure 8.7: User confirmation box.

To complete the task you will need to:

- remove the alert box and replace it with a confirm box with suitable text
- add a conditional statement to your function using 'if' and 'else'.

Save your file as **Chapter08_Task1_3.html**.

4 You have also been asked to replace the text box where the user enters a colour with a drop-down box where the user can select a colour.

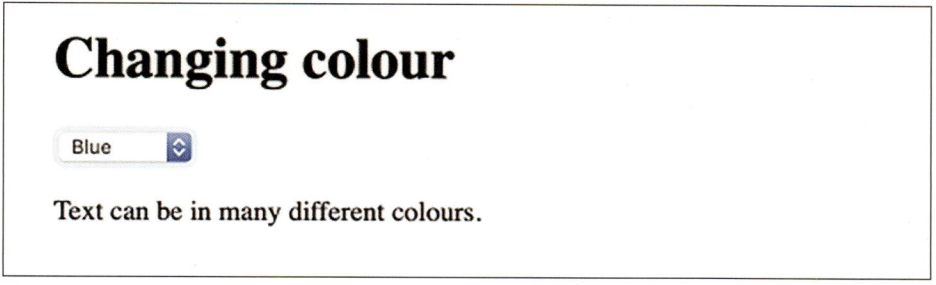

Figure 8.8: Colour selection box.

To complete the task you will need to:

- remove the input box
- add a drop-down box populated with suitable colour options and an onchange event
- change the function to obtain the selected colour from the drop-down box.

Save your file as **Chapter08_Task1_4.html**.

5 As a final improvement you have been asked to edit 1.4 so that the 'confirm' box appears three times before the colour changes.

To do this you should use a 'do…while' loop.

Save your file as **Chapter08_Task1_5.html**.

Task 2

SKILLS

This task will cover these extra skills in addition to the ones you have already used:

- creating a placeholder for an image
- importing an image depending on a user's response
- showing and hiding HTML elements
- creating functions
- using selection
- using **comparison operators**
- using mathematical operators.

KEY WORDS

comparison operator: a symbol, or symbols, that compare the two sides of the operator and the result is either true or false

1 The age of a one-year-old dog is equivalent to the age of a 12-year-old human, a two-year-old dog to a 24-year old human. Then add four years to the dog age for every year of a human after that.

The age of a one-year-old cat is equivalent to the age of a 15-year-old human, a two-year-old cat to a 24-year-old human. Then add four years to the cat age for every year of a human after that.

Your task is to create a web page where a user can find the human equivalent age of their cat or dog.

To complete the task you will need:

- a heading for the page
- a placeholder for an image of either a cat or a dog
- a button to allow the user to select either a cat or a dog
- to import an image of a cat or dog into the placeholder according to the user's choice
- a text box to allow the user to enter the animal's age
- a function to calculate the human equivalent age
- an alert to notify the user of the age.

The user should be able to click a button to select either cat or dog.

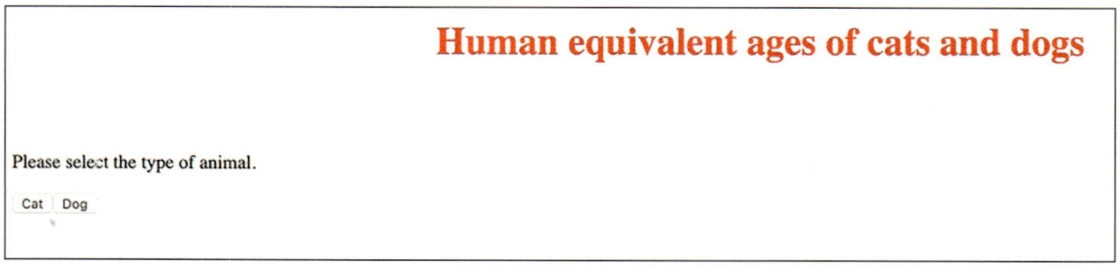

Figure 8.9: Button for selecting animal type.

After the user has selected, they should be shown an image of either a cat or a dog and a text box to enter the age.

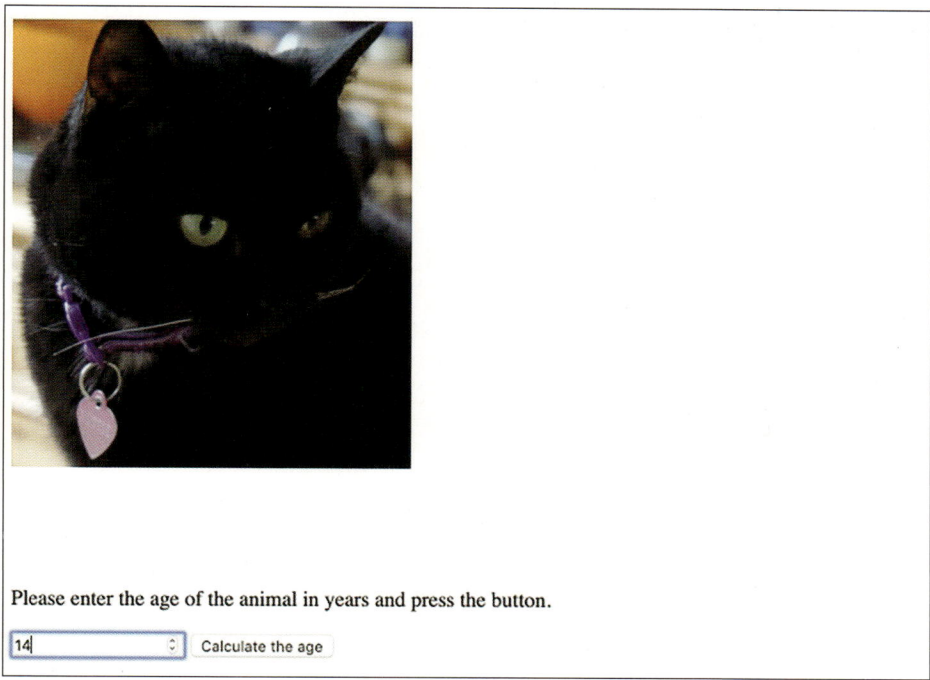

Figure 8.10: Image displayed if 'cat' is selected.

They should then be shown the human equivalent age in an alert.

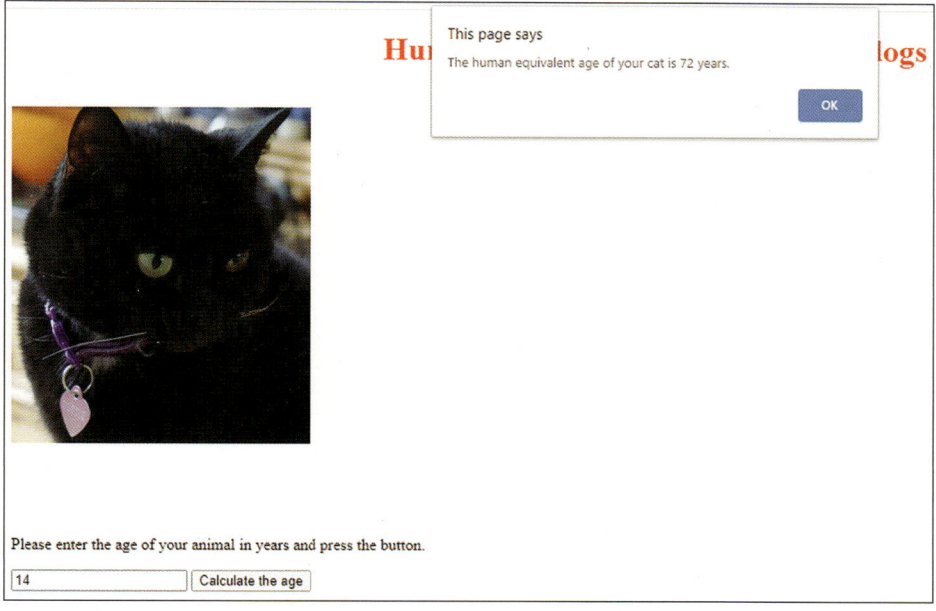

Figure 8.11: Image displayed and alert box showing human equivalent age.

Save your file as **Chapter08_Task2_1.html**.

2   You should make the following changes:

a   Make the background colour of the age entry box change to red when a user enters the age of the cat or dog.

b   To make the experience more user-friendly, it would be good if the user is asked if they want another go.

To complete the task you will need:

- a confirm to ask the user if they would like another go
- to hide elements, e.g. the image and age entry box if they click on 'OK'
- to change the background colour of the age entry box back to white
- to clear the text box where it is displayed in addition to hiding it
- give the users a goodbye message if they click 'Cancel'.

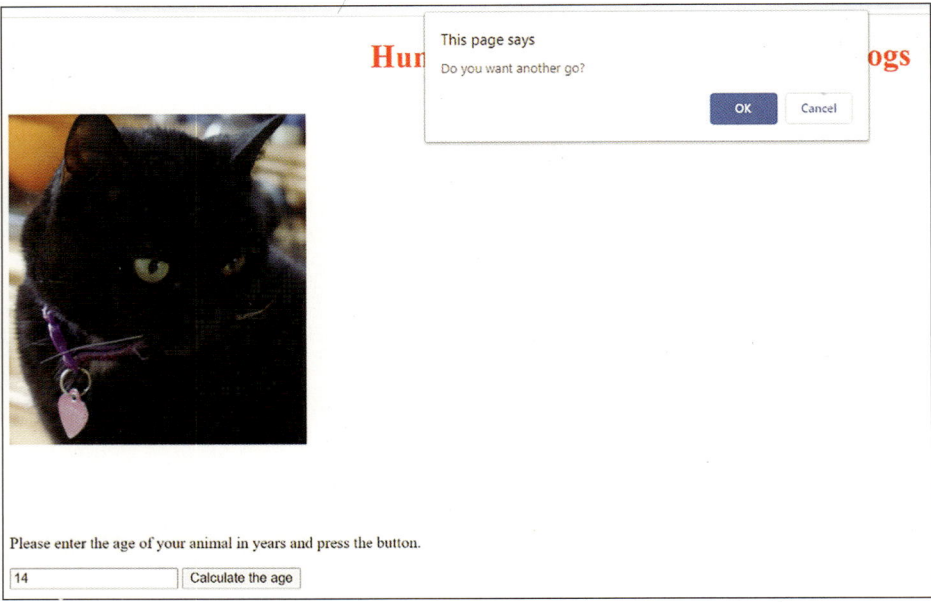

Figure 8.12: Confirmation box to select another go.

Save your file as **Chapter08_Task2_2.html**.

Task 3

You have been asked to create a web page for a taxi firm where customers can receive an estimate for their journey.

'WeDriveAnywhere' is a taxi firm with the following criteria for calculating the cost of a journey.

Between 8 a.m. to 8 p.m., the following rules apply:

- $3 for the first mile and $2 for every additional mile

- if there are more than four passengers, there is a charge of $2 for each additional passenger.

Between 8 p.m. and 8 a.m., the overall charge is doubled.

To complete the task you will need:

- a heading and a brief explanation for the customers

- text boxes where customers can enter distance, number of passengers and time

- a function to check that these have all been entered before the cost is calculated (this is called validation)

- a function to calculate the cost, with a method to determine whether the time is before 8 a.m. or after 8 p.m.

Some suggestions are shown in Figures 8.13 to 8.16.

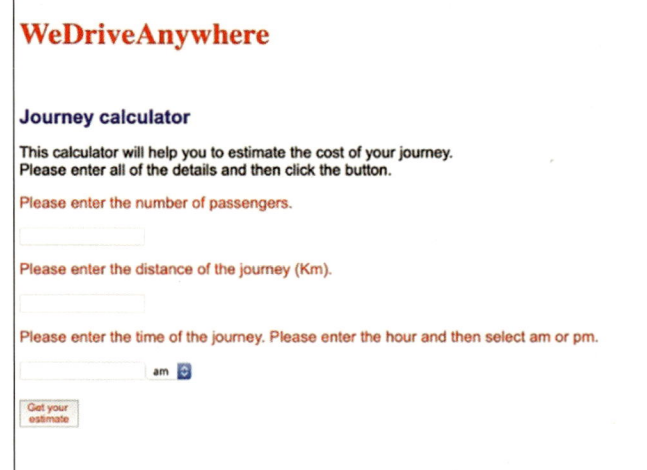

Figure 8.13: Opening page.

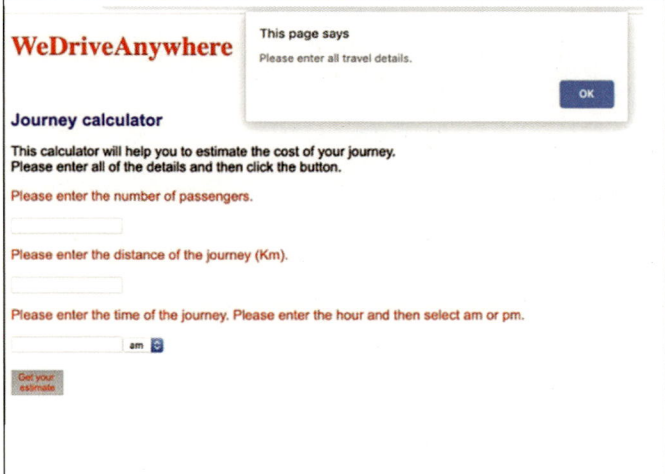

Figure 8.14: Validation to check that entries have been made.

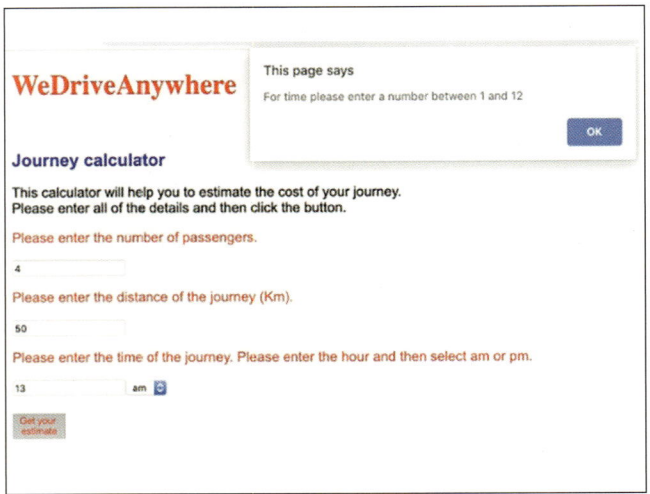

Figure 8.15: Validation to check that time entry is in the range of 1 to 12.

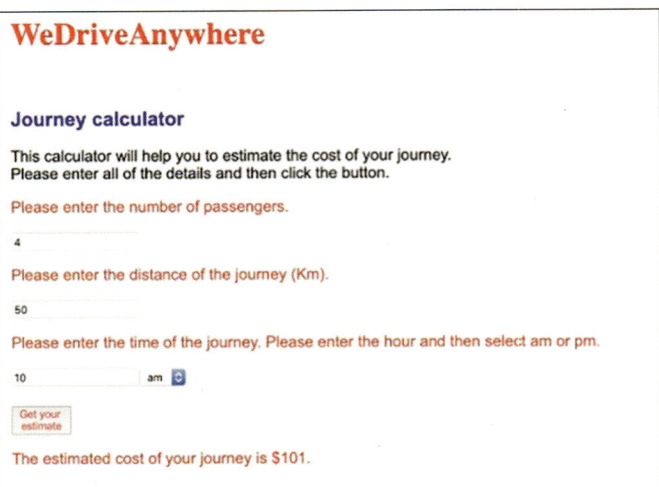

Figure 8.16: Cost estimation message for customers.

Save your file as **Chapter08_Task3.html**.

Task 4

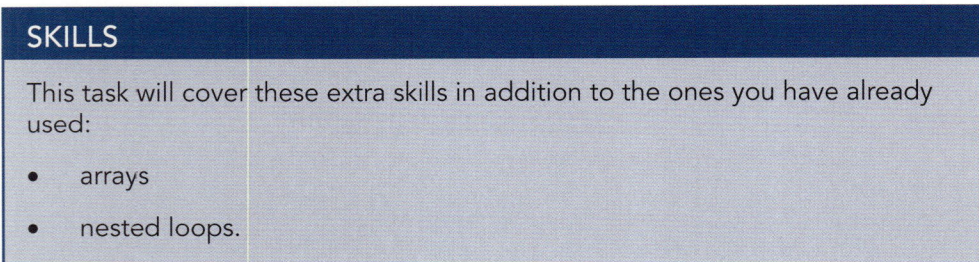

SKILLS

This task will cover these extra skills in addition to the ones you have already used:

- arrays
- nested loops.

This task requires you to solve a problem.

A student has a book containing 411 pages.

She read a certain number of pages on the first day and created a rule to work out how many pages she had to read on each succeeding day.

She decided that the number of pages to be read on the next day should be equal to the square of the sum of the digits of the page she ended at.

For example, if she ended on page 36, then she should read 81 pages on the next day as this is the square of 6 + 3.

She found that on the sixth day, the number of pages she had set herself to read took her exactly to the final page of the book.

To complete the task you will need:
- a heading and an explanation of the problem
- a button to start the calculation
- nested loops

- paragraphs to display how many pages should be read each day in addition to the total that have been read at the end of each day.

A suggestion is shown in Figure 8.17.

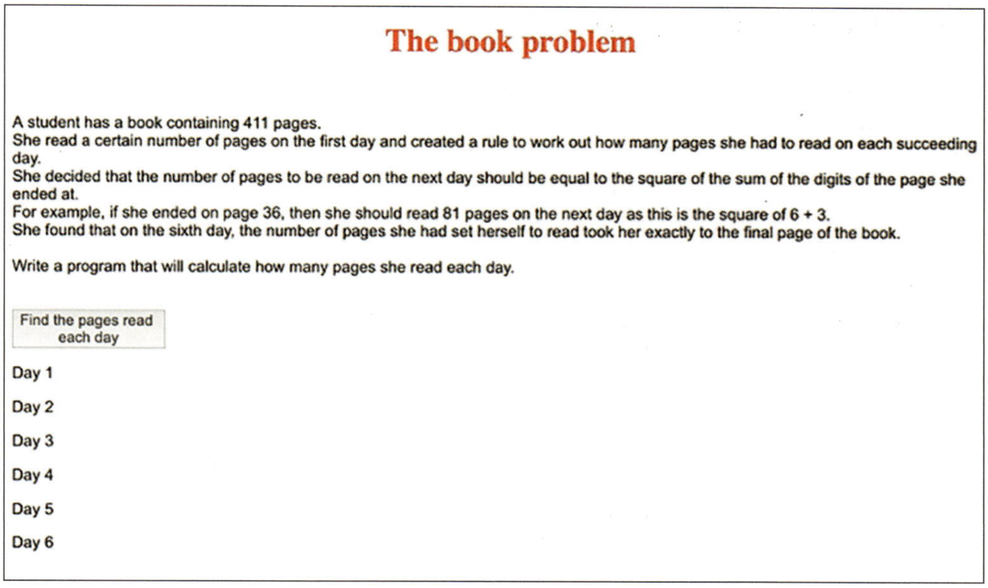
The book problem

A student has a book containing 411 pages.
She read a certain number of pages on the first day and created a rule to work out how many pages she had to read on each succeeding day.
She decided that the number of pages to be read on the next day should be equal to the square of the sum of the digits of the page she ended at.
For example, if she ended on page 36, then she should read 81 pages on the next day as this is the square of 6 + 3.
She found that on the sixth day, the number of pages she had set herself to read took her exactly to the final page of the book.

Write a program that will calculate how many pages she read each day.

Find the pages read each day

Day 1

Day 2

Day 3

Day 4

Day 5

Day 6

Figure 8.17: Opening page for users.

Save your file as **Chapter08_Task4.html**.

TIP

You could try different numbers of pages to read on day 1, i.e. 1, 2, 3 and so on. For each one apply the rules to calculate how many pages would be read each day and the total read. You can get the loop to stop when the total read is equal to 411.

Task 5

SKILLS

This task will cover this extra skill in addition to the ones you have already used:

- for/in loop.

A teacher has asked you to create an online test with the following specification:

- There should be five questions.
- At the end, the students should be shown their responses, the correct answers and finally their mark out of five.

To complete the task you will need:

- arrays to store the student responses and the correct answers
- a paragraph to display the student responses and correct answers.
- a for/in loop to traverse the arrays
- comparison operators to compare the student responses with the correct answers
- a variable to keep track of the score

Some suggestions are shown in Figures 8.18 to 8.20.

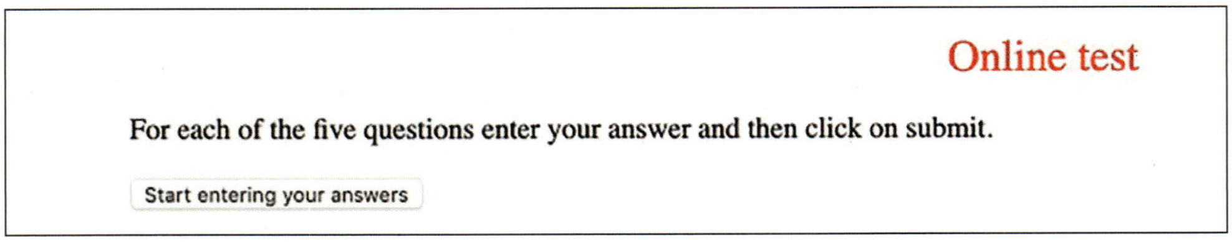
Online test
For each of the five questions enter your answer and then click on submit.
Start entering your answers

Figure 8.18: Opening page for users.

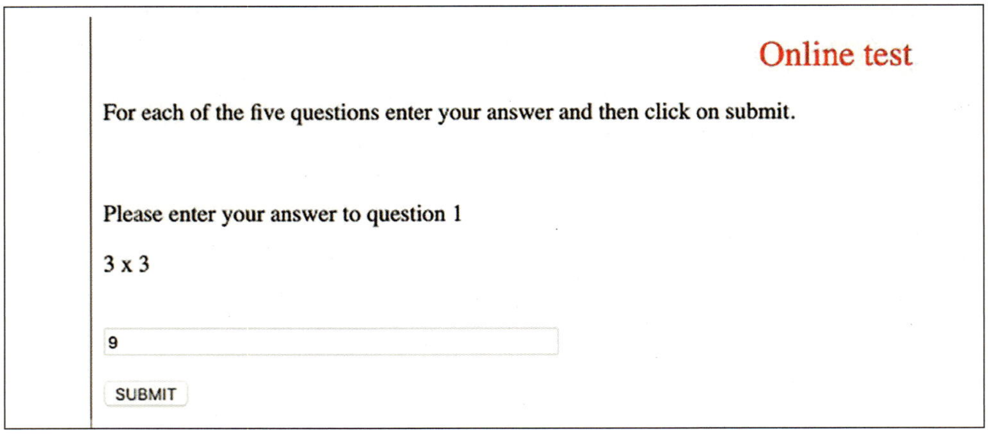
Online test
For each of the five questions enter your answer and then click on submit.
Please enter your answer to question 1
3 x 3
9
SUBMIT

Figure 8.19: First question displayed with entry box and submit button.

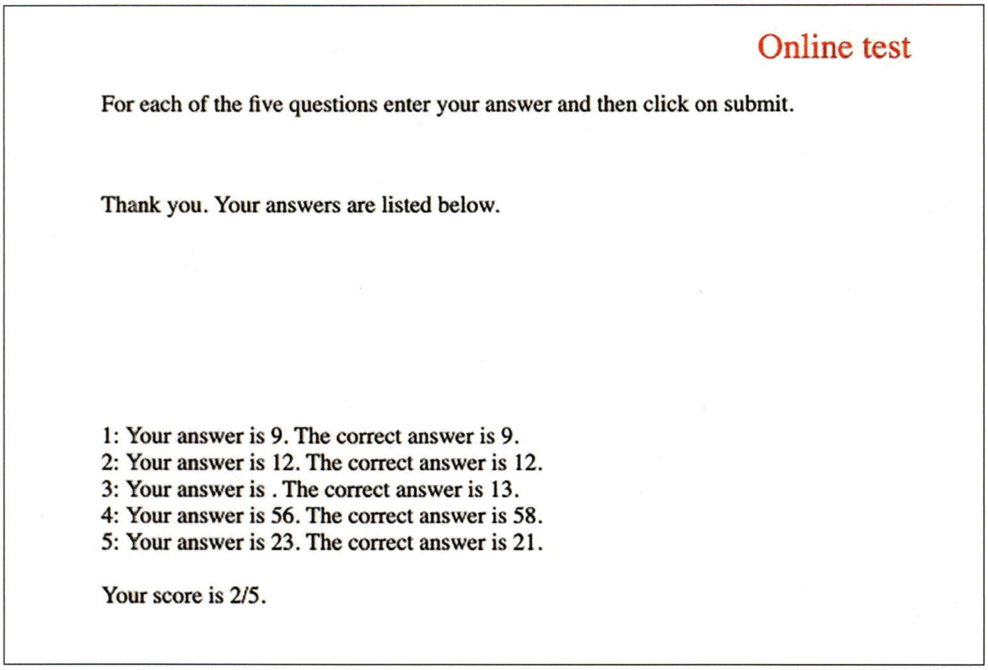
Online test
For each of the five questions enter your answer and then click on submit.
Thank you. Your answers are listed below.
1: Your answer is 9. The correct answer is 9.
2: Your answer is 12. The correct answer is 12.
3: Your answer is . The correct answer is 13.
4: Your answer is 56. The correct answer is 58.
5: Your answer is 23. The correct answer is 21.
Your score is 2/5.

Figure 8.20: Report showing the total score, and user and correct answers.

Save your file as **Chapter08_Task5.html**.

> **TIP**
>
> When you declare an array, you do not need to stipulate its length as there is a command to add an item to the end of an array.

> **TIP**
>
> Use the `<br>` tag for a line break.

Task 6

SKILLS

This task will cover these extra skills in addition to the ones you have already used:

- using radio buttons to input
- using check boxes to input.

A pizza shop offers the following ingredients for buyers to design their own pizza:

Bases

Small $2

Medium $4

Large $6

All bases are supplied with tomato sauce and cheese.

Toppings

Olives, Chicken, Cajun chicken, Red peppers, Pineapple, Tuna

Mushrooms, Sweetcorn, Onion, Jalapenos, Chillies, Cheese

All toppings cost 90c and a customer can order as many as they want.

The customer should be asked if they will collect or want it delivered. There is a 10% discount if they collect it themselves.

Your task is to design and create a program that will allow a customer to:

- select and order a base plus all of the toppings they want. When the user moves their mouse pointer over the bases a message should appear notifying them of the cost of each choice. When the user move the mouse pointer away from the bases, the message should disappear
- select collection or delivery.

When they have finished ordering, the customer should be informed of the following:

- the type of base and all of the toppings they have ordered
- the total cost of their pizza
- the customer should then be able to place their order or cancel it.

The page should then be returned to its original state for the next customer.

To complete the task you will need:

- a heading and a logical layout – options for the customer should be hidden until they are needed
- different fonts and font styles
- radio buttons for selection of base – as only one option is allowed at a time
- check boxes for the toppings as the customer can order as many as they want
- buttons to select collection or delivery

- a paragraph showing the final cost with buttons for order or cancel
- to clear all of the radio buttons and check boxes for the next customer.

Some suggestions are shown in Figures 8.21 to 8.23.

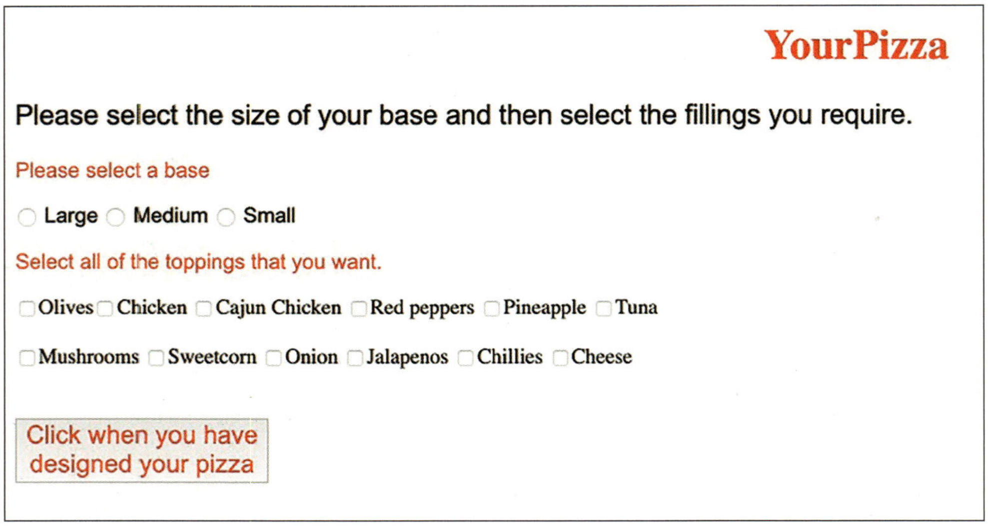

Figure 8.21: Opening page showing radio buttons and check boxes.

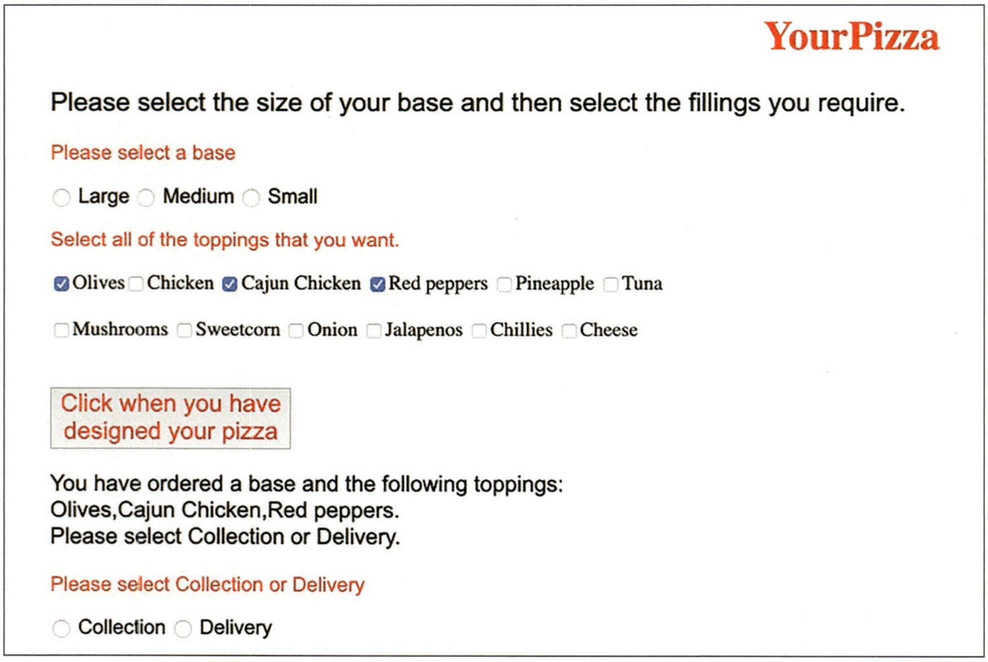

Figure 8.22: Message showing the user their order and asking for the delivery method.

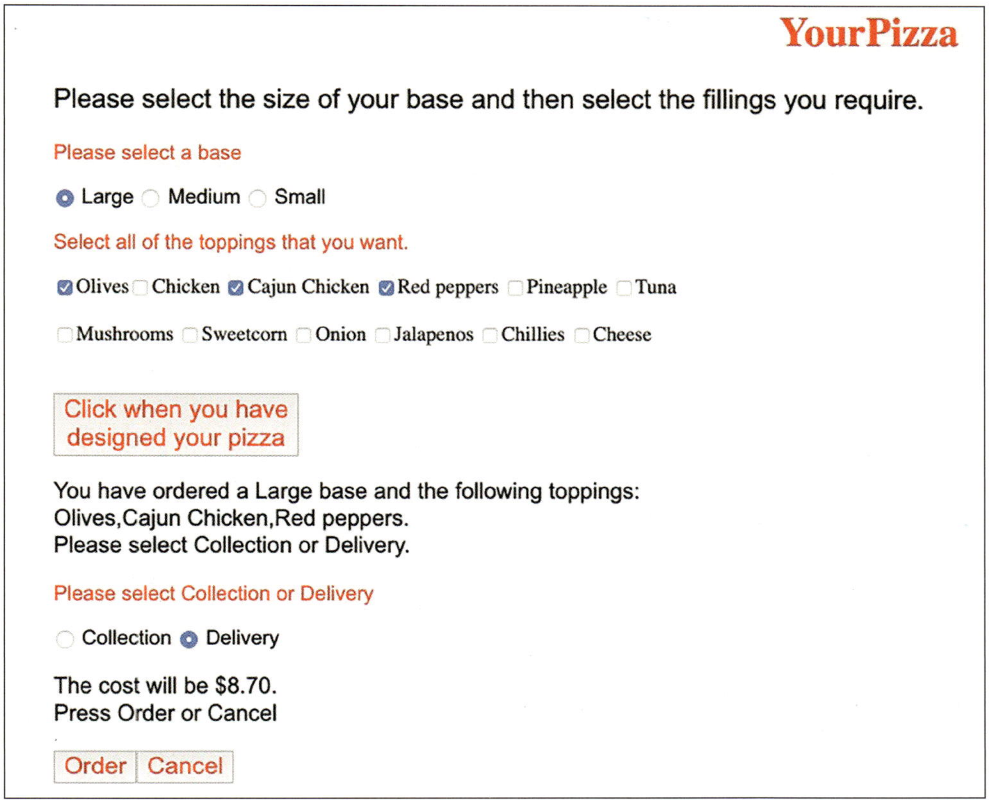

Figure 8.23: Message showing the cost with buttons to confirm or cancel the order.

Save your file as **Chapter08_Task6.html**

Task 7

> **SKILLS**
>
> This task will cover these extra skills in addition to the ones you have already used:
>
> - generating random numbers
> - using JavaScript timing events – `setTimeout()`.

Your task is to create a game involving three dice.

A player is asked if they want a new game and is given a starting amount of 500 points. At the start of each round they must give up 50 points to spin three dice.

If two dice match, they gain 200 points and 300 points if all three match.

The player can only spin the dice if they have 50 points or over remaining.

There is an added challenge. When computers simulate the throwing of dice, they are far too quick. In this game, to make it more interesting, you have to make the numbers change ten times for each dice with a time delay between each change.

To complete the task you will need to:

- ask the user if they want to start a new game

- create functions for the throwing of each dice with the time delay mentioned above

- use operators to compare the three dice to see if two or all three match

- inform the player if they have gained points, and how many

- inform the player how many points they now have

- allow the player to throw the dice again if they have at least 50 points.

Some suggestions are shown in Figures 8.24 to 8.27.

> **TIP**
>
> You will need to also set a time delay before the comparisons are made or they will be compared before all the dice have been thrown, because of the time delays on the dice throws.

Figure 8.24: Confirmation box when the page loads.

Figure 8.25: Opening page.

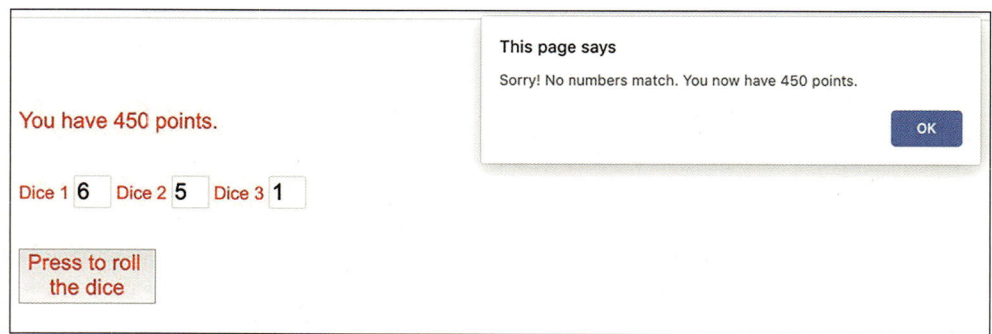

Figure 8.26: Alert showing the result when the dice do not match.

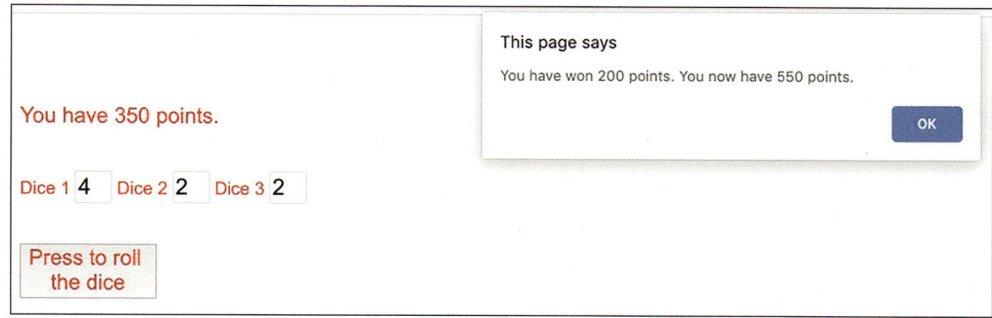

Figure 8.27: Alert showing the result when two dice match.

Save your file as **Chapter08_Task7.html**.

Task 8

SKILLS

This task will allow you to apply your skills in a more complex context.

Your task is to create a guessing game where a user has three attempts to guess a random number, between 1 and 10, generated by the computer.

The previous attempts should remain on screen, so the user does not reuse them.

The users should be informed if they guess correctly or be given the correct number after three attempts.

The user should then be asked if they want to restart the game.

To complete the task you will need:

- a button to allow the user to generate the random number

- text boxes for the user to enter their guesses – these should be hidden until they are needed

- functions to compare the guesses with the random number

- text to inform the user if they have guessed correctly or the correct number after three attempts

- a function to reset the page if the user wants to play again.

Some suggestions are shown in Figures 8.28 to 8.31.

The Guessing Game

You must click the button to generate a random number between 1 and 10. You then have 3 attempts to guess the number. Please enter a number and press the button.

Generate random number

Figure 8.28: Opening page.

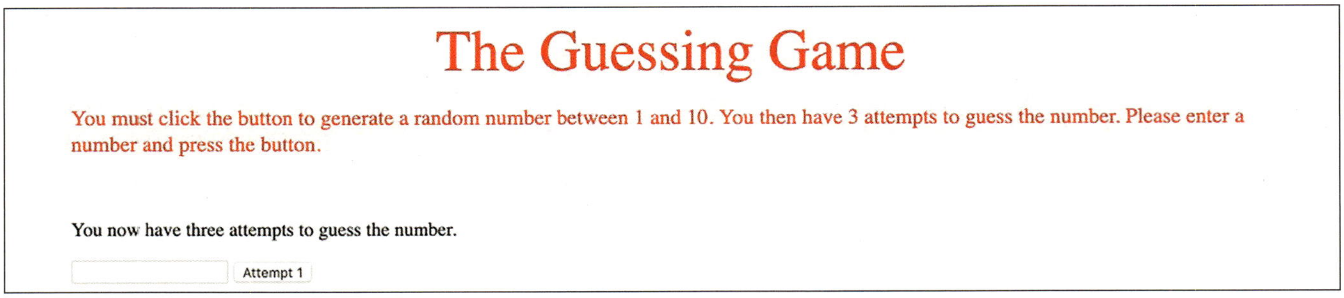

The Guessing Game

You must click the button to generate a random number between 1 and 10. You then have 3 attempts to guess the number. Please enter a number and press the button.

You now have three attempts to guess the number.

[   ] [Attempt 1]

Figure 8.29: Entry box and button for the first attempt.

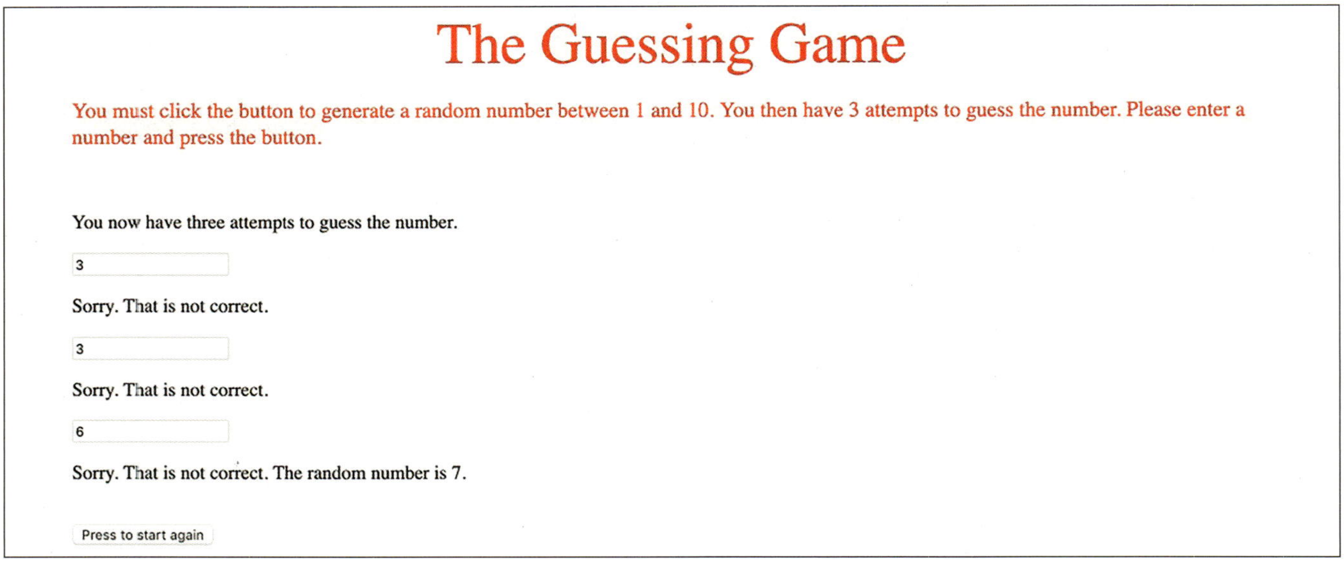

The Guessing Game

You must click the button to generate a random number between 1 and 10. You then have 3 attempts to guess the number. Please enter a number and press the button.

You now have three attempts to guess the number.

[3   ]

Sorry. That is not correct.

[3   ]

Sorry. That is not correct.

[6   ]

Sorry. That is not correct. The random number is 7.

[Press to start again]

Figure 8.30: User message when unsuccessful with the restart button.

The Guessing Game

You must click the button to generate a random number between 1 and 10. You then have 3 attempts to guess the number. Please enter a number and press the button.

You now have three attempts to guess the number.

[3   ]

Sorry. That is not correct.

[6   ]

Well done! You have guessed it at the second attempt

[Press to start again]

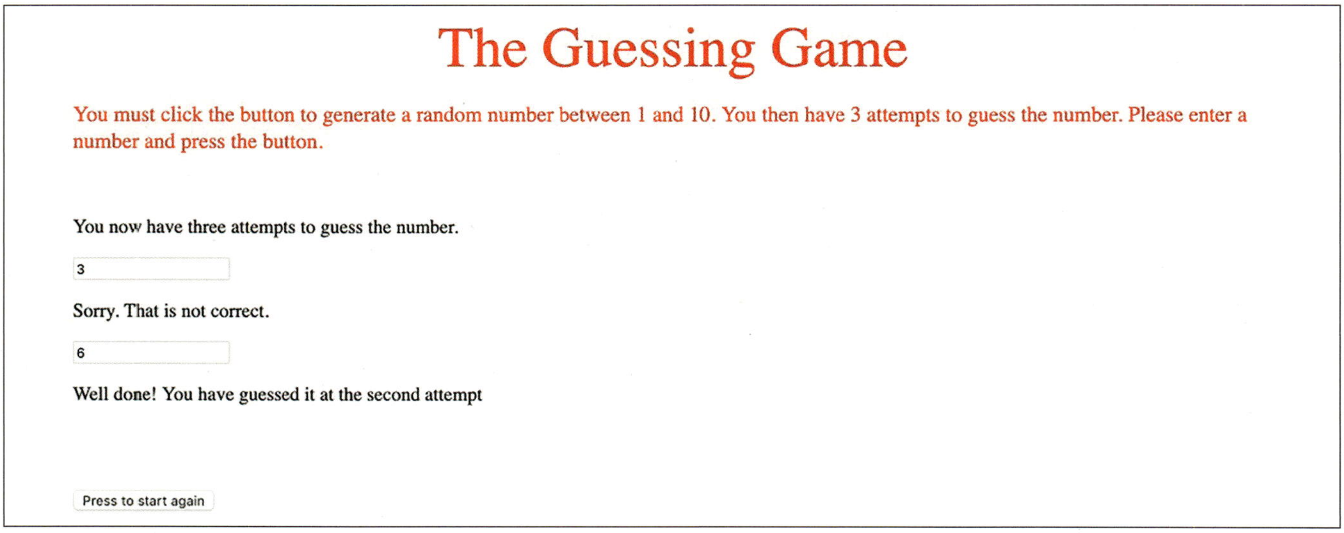

Figure 8.31: User message when successful with the restart button.

Save your file as **Chapter08_Task8.html**.

Task 9

'Caesar's Cipher' is a method of encoding and decoding text named after Julius Caesar.

It involves shifting letters of the alphabet to the left or right a set number of places. The number of spaces is called the KEY.

If the alphabet is shifted two characters to the right, every letter A would become a 'Y'. If it was shifted two characters to the left, every A would become a 'C'.

Your simulation should use a right shift as shown here.

A	B	C	D	E	F	G	H	I	J	K	L	M	N	O	P	Q	R	S	T	U	V	W	X	Y	Z
Y	Z	A	B	C	D	E	F	G	H	I	J	K	L	M	N	O	P	Q	R	S	T	U	V	W	X

To complete the task you will need:

- an explanation of what the user has to do
- text boxes for the user to enter the text, the key and whether they want to encrypt or decrypt
- validation – to check that all these have been entered
- validation – to check that the key is no greater than 25
- a function to carry out the encryption/decryption
- to allow punctuation marks to be displayed unchanged
- a paragraph where the result is displayed
- the ability to reset the page to carry out another encryption / decryption.

Some suggestions are shown in Figures 8.32 to 8.35.

TIP

An array could be used to store the letters of the alphabet (one for upper case and one for lower case) so that the new letters can be found according to the key. Instead of using the actual letters, you could use their ASCII codes. If you do not know about these, carry out some research.

Caesar's Cipher

This simple cipher is named after Julius Caesar. It involves shifting letters of the alphabet to the left or right a set number of places. The number of places is called the **KEY**.
If the alphabet is shifted 2 characters to the right every letter A would become a 'Y' and if it was shifted 2 characters to the left every A would become a 'C'.
This simulation will use a right shift.

Please enter the message, the key and whether you want to encrypt or decrypt.

Please enter the message:

Please enter the key (1 to 25):

Please enter E for encrypt or D for decrypt:

Start

Figure 8.32: Opening page.

Figure 8.33: Alert if key is not inserted.

Figure 8.34: Alert showing the error message if the number is not in the range of 1 to 25.

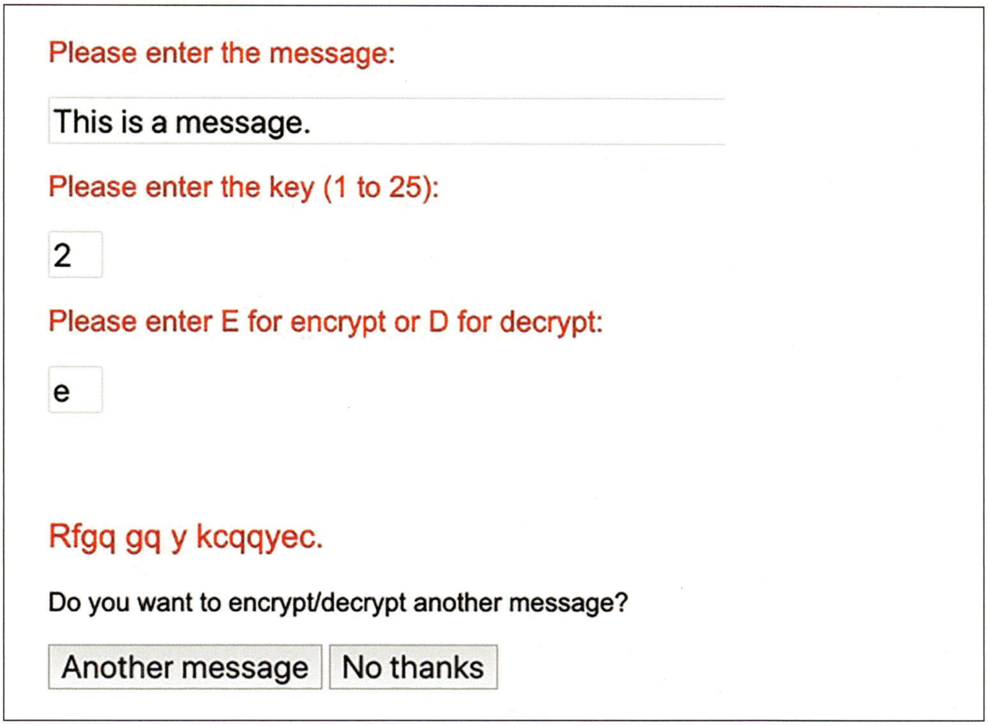

Please enter the message:

This is a message.

Please enter the key (1 to 25):

2

Please enter E for encrypt or D for decrypt:

e

Rfgq gq y kcqqyec.

Do you want to encrypt/decrypt another message?

Another message | No thanks

Figure 8.35: Encrypted message with boxes to encrypt another message or finish.

Save your file as **Chapter08_Task9.html**.

REFLECTION

- When you started to solve the problems, did you think about them first and produce possible solutions and programs on paper, or did you use the computer straight away?

- As you were designing the web page and programming using JavaScript, did you test to make sure that each section worked correctly as you created it, or did you wait until you had completed everything before you carried out any tests?

- Did you research JavaScript online when you needed to use commands and statements for the first time?

SUMMARY CHECKLIST

- [ ] I can add interactivity to web pages.
- [ ] I can change HTML content and styles.
- [ ] I can show/hide HTML elements.
- [ ] I can display data in different ways.
- [ ] I can react to common HTML events.
- [ ] I can provide user interaction.
- [ ] I can create statements.
- [ ] I can use JavaScript loops for iterative methods.
- [ ] I can create functions.
- [ ] I can use JavaScript timing events.
- [ ] I can add comments to annotate and explain code.
- [ ] I can analyse a problem and provide a solution.
- [ ] I can carry out validation on the user input.
- [ ] I can test all my programs to ensure that they function correctly.

> Glossary

Absolute cell reference: A cell reference that does not change when it is copied into other cells, usually by placing a $ sign before the cell reference.

Algorithm: A set of instructions or steps to be followed to achieve a certain outcome.

Animation: A series of images that are played one after another to simulate movement.

Array: A data structure that can store multiple items under one identifier; the items are of the same type.

Bitmap: An image made up of small squares, called pixels; each individual pixel can only be one colour.

Calculated field: An arithmetic calculation on a field from the data source

Cell: A single unit/rectangle of a spreadsheet formed at the intersection of a column and a row where data can be positioned. Its reference (name/address) is based on its column and row.

Clip: A short piece of a video or audio file.

Comparison operator: A symbol, or symbols, that compare the two sides of the operator and the result is either true or false.

Coordinates: The position (x and y) of an object on the stage.

Count-controlled loop: A loop where you know the number of times it will run.

Database: A structured method of storing data.

Embedding: Importing data from a data source so that any changes to the data source are shown in the new document.

Entity: A set of data about one thing (person, place, object or event).

Field (databases): (A common word for an attribute) A category of information about an entity stored in a database table, for example, product name, product number, ISBN code.

Field (mail merge): A category of information from the data source.

Filter (mail merge): Selecting records from the source file based on conditions.

Flowchart: A set of symbols put together with commands that are followed to solve a problem.

Formula: A mathematical calculation using +, −, × or ÷.

Frame : A single image in a video file.

Function (algorithms): A separate piece of code that has an identifier and performs a task, it can be called from elsewhere in the code and can return a value.

Function (Spreadsheets): A ready-made formula representing a complex calculation.

Global variable: A variable declared outside a function; it can be accessed anywhere within the code.

Goal seek: Looking to see what a variable needs to change to for a goal in terms of output to be achieved.

Input: Putting data into an algorithm.

Iteration : A loop to repeat a section of code for a fixed number of times or until a required outcome is achieved.

Key frame: A location on a timeline which marks a frame that has a change in the animation, for example, a drawing has changed, or the start or end of a tween.

Layer (Animation): An object or image given its own timeline for independent manipulation.

Layer (Graphics): A 'surface' onto which an image or object is placed; each object is placed on a separate layer and they are stacked on top of each other (as though on different pieces of paper).

Loop: Code that is repeated.

Macro: A set of instructions that can be completed all at once.

Mail merge: The automatic addition of data, such as names and addresses, from a source file into a master document, such as a letter.

Master document: The main document into which the data will be merged.

Nested loops : One construct that is inside another construct.

Object (Animation): An image, or combination of images, that is manipulated as one item.

Opacity: The lack of transparency of an image; at 0% opacity the image is fully transparent

Output: Displaying data from an algorithm to the user.

Parameter (Algorithms): A piece of data that is sent to a subroutine.

Parameter (Databases): The search criteria stipulated in a query.

Pixel: A small square of one colour; these are combined to create a bitmap image.

Primary key: A field that contains the unique identifier for a record.

Procedure: A type of subroutine that does not return a value to the main program.

Process: An action performed to some data to make a change.

Pseudocode: A language that is used to display an algorithm.

Query: A request for data from a table or combination of tables.

Range: A selection of cells.

Record: Consists of all the fields about an individual instance of an entity in a database, for example all the details about one student.

Relationship: The way in which two entities in two different tables are connected.

Relative cell reference: A cell reference that changes when it is copied into other cells.

Reverberation: The number of echoes of the sound and the way that they decay. These can be natural caused by the room in which the sound is played or introduced by editing software.

RGB: Red/Green/Blue colour system; all colours are a combination of quantities of red, green and blue.

Selection: Use of a conditional statement to decide a course of action or which section of code to run.

Source file: The file containing the data that will be merged into the master document.

Splice: Join together two or more sound or video clips.

Spreadsheet: Software that can organise, analyse and manipulate data organised in a grid of rows and columns.

Stage: The area where the animation takes place; to be within the animation, the object must be on the stage.

Subroutine: A set of instructions that have an identifier and that are independent from the code; it is called from another part of the program and returns control when it is finished.

Table: A collection of related data, organised in rows and columns (for example, about people, places, objects or events).

Timeline: The place that controls the order the frames are run, the positioning of the layers and so on.

Track: A specific recording, for example, of one instrument or voice. The tracks can then be edited separately and combined to play concurrently.

Transition: The method with which one video clip merges into a second clip.

Tween: (Inbe*tween*ing) An animation where the start and end points are set; the computer generates the actual images to make the animation change.

Variable: A space in the memory of a computer that has an identifier, where you can store data. This data can be changed.

Vector: An image that uses geometric points and shapes; calculations are used to draw the image.

What-if analysis: Experimenting with changing variables to see what would happen to the output if those variables changed.